Books for All Learners Lessons
On-Level

Grade 3

Harcourt
SCHOOL PUBLISHERS

Visit *The Learning Site!*
www.harcourtschool.com

Printed in the United States of America

ISBN 0-15-356699-X

Contents

Giggle and Snort

by Anne Miranda

Poetry

SUMMARY Giggle and Snort, two young pigs, pride themselves on not listening in class and refusing to follow rules. When Giggle and Snort visit a ranch, they learn the hard way what happens when they don't pay attention and follow the rules of safety.

- **Focus Skill:** Decode Long Words
- **Focus Strategy:** Use Context to Confirm Meaning
- **Vocabulary:** department, obey, commands, audience, expression, accident, noticed

Differentiated Instruction

Paired Reading Organize students in pairs. Have them read the selection aloud to each other, alternating with each page. After students have finished reading, have them discuss in pairs a part of the story they found interesting.

Prepare to Read

Preview/Set Purpose Have students examine the cover and title. Have them preview the book by looking at the illustrations. Ask students which two pigs seem to be the main characters in the story. Point out that this is a book of humorous poetry. Ask them what they think this book will be about. Have students volunteer any horseback-riding experiences they might have.

Focus Skill

Decode Long Words

Teach/Model Tell students that good readers break a long word into smaller parts to help them understand the meaning of the word. Sometimes the word is a compound word, or a word made of two or more smaller words. Write the word *horseback*. Say, *When I see the word* horseback, *I see two smaller words that I know*: horse *and* back. *I think* horseback *means the back of a horse. I can look up this word in a dictionary to confirm its meaning.*

Practice/Apply Tell partners to look for words they already know in longer words as they read *Giggle and Snort*.

Focus Strategy

Use Context to Confirm Meaning

Teach/Model Tell students that good readers look for the context, or meaning, of the words and sentences around a word they find confusing. The context can give clues to the meaning of an unfamiliar word. Read the first stanza of *Giggle and Snort* on page 2. Say, *I'm not sure about the meaning of* horseback. *The phrase* horseback riding *might be a clue. I know that people usually ride horses with a saddle. Maybe* horseback riding *means riding directly on the horse's back.*

Practice/Apply Tell partners to use context to confirm the meanings of unfamilar words and phrases as they read the book.

Build Vocabulary

Help students use context to understand key vocabulary. Have them answer each question by reading the words and sentences that surround the word.

department (p. 5)	What three word parts make up the word *department*? Name an example of a department. **de + part + ment; fire department, police department**
obey (p. 5)	Give one example of a law that we must *obey*. **Possible response: Stop at a stop sign.**
commands (p. 5)	What *commands* might a soldier be given? **Possible response: March! Halt!**
audience (p. 7)	What does the *audience* do to encourage an actor? **Possible response: clap, cheer**
expression (p. 6)	What might your *expression* look like if you eat a lemon?**Possible response: squinted eyes, pursed lips, tongue sticking out, wrinkled nose**
accident (p. 6)	"I didn't mean to sleep late this morning. It was an *accident*!" What words give you context to figure out what *accident* means? **I didn't mean to**
noticed (not in text)	If you *noticed* a mud puddle in front of you, what would you do? **Possible response: jump in it, walk around it**

CM 7 **Independent Practice:** Vocabulary Practice page 7

Read and Comprehend

Monitor Comprehension

Guide students as they read the text, using the following questions:

Pages

2–3 What do you think will happen in the story after you learn Giggle and Snort do not listen to the horse-safety rules? **Possible response: They will get hurt.** PREDICT OUTCOMES

4–6 "We never obey. That's not our *department*." Think about what you know about the meaning of *department*. What is the meaning in this context? **Possible response: part of overall behavior** USE CONTEXT TO CONFIRM MEANING (Focus Strategy)

7–8 What did Giggle and Snort do when the class went into the stable? **They went outside.** STORY EVENTS

9–12 What do Giggle and Snort look like when Big Roan starts running? **a strange piggy kite** FIGURATIVE LANGUAGE

14–16 What lesson do Giggle and Snort learn? **to listen to rules** SUMMARIZE

Think and Respond

Have students answer the questions on the inside back cover of the book.

1. **Possible response: This story is special because it was written in the form of a poem. The rhyme makes the story easy and fun to read.**
2. ***Department* has three syllables.**
3. **They talk about safety rules because they don't want the children to get hurt.**
4. **Possible response: I learned that riding horses is not easy.**
5. **The talks are the same because they both go over the safety rules. They are different because Miss Flower has a picture and Gus has an actual horse.**
6. **Possible response: I don't think Giggle and Snort were very smart at first, but I liked them better after they learned that rules are important.**

SOCIAL STUDIES

Social Studies Connection Write the beginnings of some well-known safety rules. For example, "Look both ways before you ____" or "Wear a helmet when you ____." Have students name possible responses to complete the rules.

School-Home Connection Have students write a list of safety rules they follow regulary.

CM 8 **Independent Practice:** Comprehension Practice page 8

Write

Writer's Craft: Voice

Write a Descriptive Paragraph Read page 12 of *Giggle and Snort* with students. Say, *The writer shows how funny and exciting this scene is with exclamation points and descriptive words like* plopped. Tell students that good writers use punctuation and descriptive words or phrases to express their feelings, or voice. Display a web like the one below. In the circle write *funny, exciting.* On one leg write *many exclamation points*; on another write the word *plopped.* Work with students to find other words on page 12 that reflect the humorous voice and write them on the third leg of the web (*strange piggy kite*). Encourage students to use a similar web to help them express their voice in a descriptive paragraph about something funny or exciting that has happened to them.

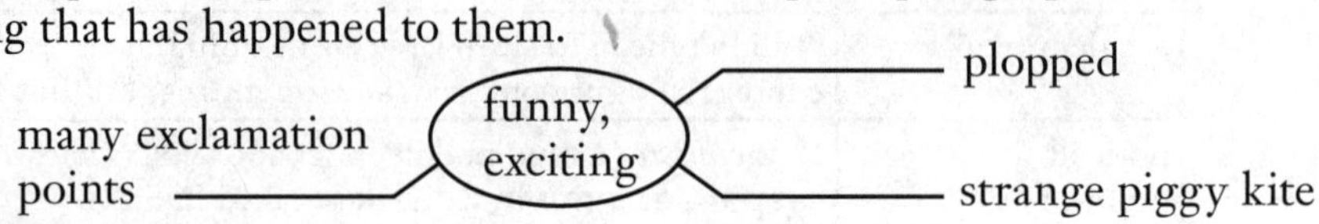

Fluency

Oral Reading Fluency

CM 9

Assessment Use Oral Reading Fluency Practice page 9 to assess students' oral reading fluency.

Name ____________________ Giggle and Snort Vocabulary

▶ **Write the meaning of the word in dark print. Look for context clues.**

1. When I looked up at her face, I was shocked by the **expression** I saw on it.
   An **expression** is the look on someone's **face**.
2. Ed did not want to do what Sue asked, but finally he **obeyed**.
   Obey means **do what is asked**.
3. The **audience** watching the show clapped loudly.
   An **audience** is a group of people who **watch a show**.
4. The general shouted out **commands**, or orders, and the soldiers marched ahead.
   Commands are **orders**.

▶ **Describe a funny incident. Use the words *accident, noticed,* and *department.***

Possible response: I can't believe what happened in the toy department of the store yesterday! A girl was throwing a ball to her friend and knocked over a whole shelf of toys by accident! She looked to see if anyone had noticed, but she didn't see me laughing.

© Harcourt

Books for All Learners Lessons, Grade 3 7

▲ Vocabulary Practice p. 7

Name ____________________ Giggle and Snort Comprehension

▶ **Circle the letter of the best answer for each question below.**

1. How would you break the word *performance* into word parts?
   A per + for + mance
   B perf + or + mance
   Ⓒ per + form + ance
   D perf + orm + ance
2. How did Giggle and Snort feel while they were riding Big Roan?
   F full of energy
   G tired and sick
   H proud of themselves
   Ⓙ afraid and nervous
3. What is the theme of this story?
   A Horses can be dangerous.
   B Laughter is the best medicine.
   C Hold on tight while riding a horse.
   Ⓓ Rules are important to follow.

▶ **Read this sentence from the story and answer the question below.**

Gus jumped on the back of his shiny brown steed.

What is a steed? How do the other words in the sentence or story help you find the meaning of the word?

Possible response: A steed is a horse. The rest of the story talks about horseback riding. In this sentence Gus jumps on the back of the steed, so a steed is another word for a horse.

© Harcourt

8 Books for All Learners Lessons, Grade 3

▲ Comprehension Practice p. 8

Name ______________________________

▶ **Write the meaning of the word in dark print. Look for context clues.**

1. When I looked up at her face, I was shocked by the **expression** I saw on it.

   An **expression** is the look on someone's ____________________.

2. Ed did not want to do what Sue asked, but finally he **obeyed**.

   Obey means ____________________.

3. The **audience** watching the show clapped loudly.

   An **audience** is a group of people who ____________________.

4. The general shouted out **commands**, or orders, and the soldiers marched ahead.

   Commands are ____________________.

▶ **Describe a funny incident. Use the words *accident*, *noticed*, and *department*.**

____

____

____

____

____

____

Name ______________________________

▶ Circle the letter of the best answer for each question below.

1. How would you break the word *performance* into word parts?

A per + for + mance
B perf + or + mance
C per + form + ance
D perf + orm + ance

2. How did Giggle and Snort feel while they were riding Big Roan?

F full of energy
G tired and sick
H proud of themselves
J afraid and nervous

3. What is the theme of this story?

A Horses can be dangerous.
B Laughter is the best medicine.
C Hold on tight while riding a horse.
D Rules are important to follow.

▶ Read this sentence from the story and answer the question below.

Gus jumped on the back
of his shiny brown steed.

What is a steed? How do the other words in the sentence or story help you find the meaning of the word?

____

____

____

____

Name ______________________________

Big Roan took off running. The two pigs held tight. **10**
They streamed out behind like a strange piggy kite. **19**
Snort went to the left. Giggle went to the right. **29**
Big Roan must have thought they were having a fight! **39**
Giggle flew up. Then he plopped down on Snort. **48**
"Get off of me!" his pal yelled. "I don't like this sport!" **60**
They stood on the saddle and tried changing places. **69**
No one could see the fear on their faces. **78**
They rode round the ring with Big Roan in high gear. **89**
All the pigs watching let out a big cheer. **98**
Then Snort grabbed the reins and pulled back with one arm. **109**
Giggle yelled "WHOA!" and it worked like a charm. **118**
Big Roan came to a stop. He was happy to rest. **129**
It was quite a performance –maybe his best! **137**
All the students applauded the piglets' great trick. **145**
But Giggle and Snort were both feeling quite sick. **154**
The cowhand rode over and shouted with glee. **162**

FLUENCY SCORE
Total Words Read Per Minute __________
Number of Errors — __________
Number of Words Read Correctly __________

Fiction

Mischa Learns English

by Manya Weg

SUMMARY Mischa and his family have just arrived in California from Russia. Though he doesn't speak English, Mischa quickly learns the language from his new friends. When Mischa's language skills are put to the test, he realizes how much English he has learned.

- **Focus Skill:** Narrative Elements
- **Focus Strategy:** Make and Confirm Predictions
- **Vocabulary:** stubborn, grumble, mumbled, exploded, darted, streak, languages

Differentiated Instruction

Paired Reading Organize students in pairs. Have them read the selection aloud to each other, alternating with each page. After students have finished reading, have them discuss in pairs a part of the story they found interesting.

Prepare to Read

Preview/Set Purpose Have students look at the illustrations and the title of the book. Have students tell what they think the book will be about. Ask them what country they think Mischa might be from. Tell students that this is a fiction story, in which the characters and the events are made-up.

Focus Skill

Narrative Elements

Teach/Model Draw a story chart similar to the one below. Tell students that good readers think about narrative elements such as characters, setting, and plot to help them understand what they read. Explain that characters are the people a story is about; setting is the time and place in which a story happens; and plot tells the events that happen in a story. Tell students that a plot often includes a problem and a solution. Have students read page 2. Say, *This story is about a boy named Mischa. He is one of the characters. Mischa lives with his family in California, where this story takes place. Ricky, another character in the story, tries to retrieve his soccer ball from tree branches. This is one of the events in the story. All of these are narrative elements in* Mischa Learns English.

Characters	Setting	Events
Mischa, Ricky	California	Ricky is trying to get his soccer ball that is stuck in a tree.

Practice/Apply Have partners create story charts and add other characters and events after they have read the story.

Focus Strategy

Make and Confirm Predictions

Teach/Model Tell students that good readers make and confirm predictions as they read a story. Explain that a prediction is a guess and learning whether or not you are correct is called confirming a prediction. Point out that pausing to make and confirm predictions will help students better understand the problem and solution in the plot. Read page 5 with students. Say, *This is Mischa's first day at school. Since he speaks very little English, I predict that his teacher will find a way to help him understand what others are saying. I will read ahead to see if I guessed right.*

Practice/Apply Tell partners to make and confirm preditions as they read the story.

Build Vocabulary

Help students use context to understand key vocabulary. Have them answer each question by reading the words and the sentences that surround the words.

stubborn (p. 4)	What word could you use to describe a *stubborn* person? **headstrong, difficult**
grumble (p. 4)	When would you *grumble*? **when you are complaining about something**
mumbled (p.12)	What is a phrase that means the same as *mumbled*? **said unclearly**
exploded (p. 3)	If something *exploded* nearby, why might you cover your ears? **because it would make a loud noise**
darted (p. 3)	If a cat *darted* outside, what did it do? **ran quickly outside**
streak (p. 3)	When might you say cars *streak*? **when they drive very fast**
languages (p. 7)	How can people communicate if they speak different *languages*? **They can point, draw pictures, and make gestures.**

CM 13 **Independent Practice:** Vocabulary Practice page 13

Read and Comprehend

Monitor Comprehension

Guide students as they read the text, using the following questions:

Pages

2–3 Why did Mischa help Ricky get the soccer ball out of the tree? **Mischa is a nice person and wanted to make friends.** MAKE INFERENCES

4–5 Do you think that Ricky will help Mischa adjust to school in America? Why or why not? **Possible response: Yes, because Ricky seems like a good friend.** MAKE PREDICTIONS (Focus Strategy)

6–9 How is Mrs. Parker described? **She smiles and giggles a lot and everyone likes her.** CHARACTERIZATION

10–12 Why doesn't Mischa talk much at lunch? **He doesn't think he knows enough English.** DRAW CONCLUSIONS

13–16 How does Mischa help Valentina? **He translates what the nurse says from English into Russian.** STORY EVENTS (Focus Skill)

Think and Respond

Have students answer the questions on the inside back cover of the book.

1. **Mischa is the "I" in the story.**
2. **Mischa helps a hurt little girl who cannot understand English.**
3. **Possible response: She wanted the students to think about the different parts of the world that their classmates came from.**
4. **Possible response: I learned that it is important to help someone who is adjusting to a new culture and language.**
5. **Mischa is the same at the beginning and end of the story because he is Russian. He is different because he has more confidence in himself.**
6. **Possible response: I would make friends and try to learn the language, like Mischa.**

Language Arts Connection Have students work in pairs to list what details would change if the story was told from Ricky's or Natalie's point of view.

School-Home Connection Have students share *Mischa Learns English* with a family member. Tell them to ask their family member if he or she has ever known someone from a different country.

CM 14 **Independent Practice:** Comprehension Practice page 14

Write

Writer's Craft: Voice

Write a Diary Entry Have students reread page 13 of *Mischa Learns English.* Say, *Mischa has been asked to translate Russian for the school nurse. Look at how the writer describes Mischa's thoughts: "I wondered whether I really would be able to help. I knew I had learned a few words, but would I really find the words I needed?" This description tells me that Mischa is worried about translating.* Explain that writers express a personal voice through the way they use language. Vivid details and descriptive language help make a writer's voice strong. Writers also use sense words to tell how something looks, tastes, smells, sounds, or feels. Have students write a diary entry about a time that they helped someone. Remind students to use words that express their thoughts and feelings so that their diary entries have a personal voice.

When I helped...

Fluency

Oral Reading Fluency

CM 15

Assessment Use Oral Reading Fluency Practice page 15 to assess students' oral reading fluency.

Name ______________________

Mischa Learns English
Vocabulary

▶ **Mark each sentence *True* or *False*. If the sentence is false, rewrite it to make it true.**

F 1. *Stubborn* describes a person who gives in to others.
Stubborn describes a person who won't give in.

T 2. *Languages* are systems people use to talk with one another.

T 3. If words are *mumbled*, they are spoken in a low, unclear way.

F 4. A *streak* is a slow movement.
A streak is a quick flash of movement.

T 5. If a bottle has *exploded*, it has burst into many tiny pieces.

F 6. To *grumble* means to complain loudly.
To grumble means to complain quietly.

F 7. If a dog *darted* across your path, it moved slowly.
If a dog *darted* across your path, it moved quickly.

▶ **Imagine you are Valentina. Using two vocabulary words, write a note to Mischa thanking him for his help.**
Possible response: Mischa, Thanks for helping me translate between two languages. Your words were not mumbled, and they helped me a lot! Your friend, Valentina.

▲ Vocabulary Practice p. 13

Name ______________________

Mischa Learns English
Comprehension

▶ **Fill in the story map with details from *Mischa Learns English.***

Setting
California
Mischa's house
Mischa's school

Characters
Mischa, his mother, Mrs. Parker, Ricky, Natalie, classmates, school nurse, Valentina

Beginning:
Problem
Mischa worries that he will not learn English or fit in.

Middle:
Main Events/Details
Mischa makes friends and begins to learn English.

Ending:
Solution to Problem
Mischa translates for Valentina.

▶ **Answer the questions below.**

How did you think Mischa would fit in at his new school? Why did you guess this? Did you confirm your prediction?
Possible response: I predicted that Mischa would make friends at his new school because it seemed like Ricky was going to help Mischa fit in. I confirmed my prediction when I read that Mischa started learning English and had friendly classmates.

▲ Comprehension Practice p. 14

Name ______________________________

▶ **Mark each sentence *True* or *False*. If the sentence is false, rewrite it to make it true.**

____ **1.** *Stubborn* describes a person who gives in to others.

____

____ **2.** *Languages* are systems people use to talk with one another.

____

____ **3.** If words are *mumbled*, they are spoken in a low, unclear way.

____

____ **4.** A *streak* is a slow movement.

____

____ **5.** If a bottle has *exploded*, it has burst into many tiny pieces.

____

____ **6.** To *grumble* means to complain loudly.

____

____ **7.** If a dog *darted* across your path, it moved slowly.

____

▶ **Imagine you are Valentina. Using two vocabulary words, write a note to Mischa thanking him for his help.**

____

____

____

Name ________________________________

▶ **Fill in the story map with details from *Mischa Learns English*.**

Setting
California
Mischa's house
Mischa's school

Characters

Beginning:
Problem

Middle:
Main Events/Details

Ending:
Solution to Problem

▶ **Answer the questions below.**

How did you think Mischa would fit in at his new school? Why did you guess this? Did you confirm your prediction?

____

____

____

____

Name ______________________________

Natalie introduced me as Mischa. I quickly whispered to her. 10
Then she introduced me as Mike. Much better, I thought to myself. 22
Mrs. Parker pulled down a map of the world and showed us the 35
United States of America. She pointed out the Pacific Ocean to the west 48
and the Atlantic Ocean to the east. America was a very big country, I 62
thought to myself. It stretched all the way from one ocean to the other. 76
Then Mrs. Parker asked me to show everyone where Russia was 87
on the map. It was on the other side of the Atlantic Ocean. Russia 101
was huge. Two big countries, and they're both mine! I thought. 112
Mrs. Parker asked Sung to point to South Korea, where his 123
family had come from. She asked other children to show on the 135
map the country their family had come from. Then some students 146
showed the different state where their families had lived before 156
they came to California. Some students had always lived in 166
California. 167

FLUENCY SCORE

Total Words Read Per Minute ____________

Number of Errors − ____________

Number of Words Read Correctly ____________

Fiction/Mystery

The Case of the Missing Case

by Anne Miranda

SUMMARY Violinist Wolfie Bach has lost his violin case and calls on the world-renowned cat detective, Miss Tourees, for help.

- **Focus Skill:** Decode Long Words
- **Focus Strategy:** Read Ahead
- **Vocabulary:** detective, case, specific, assistant, definitely, returned, positive

Differentiated Instruction

Paired Reading Organize students in pairs. Have them read the selection aloud to each other, alternating with each page. After partners have finished reading, have them discuss a part of the story they found interesting.

Prepare to Read

Preview/Set Purpose Have students examine the cover, title, and illustrations in the book. Remind students that this is a fiction book, which is a book about made-up people and events. Have students relate times when they lost something and how they found it.

Focus Skill

Decode Long Words

Teach/Model Draw a decoding chart similar to the one below. Tell students that good readers decode long words by breaking them into smaller parts. Some words are made up of a root word with an ending, as in the word *interesting*, in which the root word is *interest* and *-ing* is added on. Explain that *interesting* means "causing interest." Other long words are made up of two smaller words, as in *kettledrum*, which is made of *kettle* and *drum*. Tell students that a *kettledrum* is a drum shaped like a kettle.

Word from Text	Page	Syllables	Smaller Words	Word Meaning
interesting	2	in/ter/est/ing	interest	causing attention or interest
kettledrum	6	ket/tle/drum	kettle-drum	a drum that is shaped like a kettle

Practice/Apply Have partners create their own decoding chart and decode the following words: *peppermint*, *violinist*, *onstage*, and *curvy*.

Focus Strategy

Read Ahead

Teach/Model Tell students that good readers read ahead if they are confused or don't understand something in the text. Explain that when reading a mystery, good readers use clues to try to figure out what will happen at the end of a story. They might read ahead to find out about events that are not explained until later in the story. Have students read the first few lines of page 3. Say, *When I read that Miss Tourees has a new case, I want to know what it is about. So I read ahead to find out. I soon learn that Miss Tourees has to investigate the missing case of a famous violinist.* Next, tell students that when they see an unfamiliar word, they can try reading ahead to see if the word is used in another sentence. This might help them find the word's meaning.

Practice/Apply Tell partners to read ahead in the text to help them understand confusing passages.

Build Vocabulary

Help students use context to understand key vocabulary. Have them answer each question by reading the words and sentences that surround the word.

detective (p. 3)	What work do *detectives* do? **They investigate cases and look for evidence.**
specific (p. 4)	What is an antonym of *specific*? **general, overall**
positive (p. 5)	How do you feel if you are *positive* about something? **You are sure about it.**
case (p. 3)	When a detective gets a *case*, what does he or she try to do with it? **solve it**
returned (p. 4)	If a pet dog is *returned*, what has happened to him? **He was lost and now is being given back to his original owner.**
definitely (p. 5)	If Kate is *definitely* telling the truth, could she be lying? **No, she couldn't be lying.**
assistant (p. 5)	What smaller word can you find in the word *assistant*? What does someone's *assistant* do? **assist; helps a person in his or her work**

CM 19 **Independent Practice:** Vocabulary Practice page 19

Read and Comprehend

Monitor Comprehension

Guide students as they read the text, using the following questions:

Pages

2–4 What two smaller words make up *priceless*? **price and less.** DECODE LONG WORDS (Focus Skill)

5–7 Why does Miss Tourees want to talk to the janitor? **He can get into Wolfie's dressing room.** IMPORTANT DETAILS

8–10 How does Miss Tourees first sort the cases? **by size** SEQUENCE

11–13 What does Miss Tourees suggest that Wolfie do in the future if he loses his case? **use a name tag** PROBLEM-SOLUTION

14–16 What else is in Wolfie's case? **a bow, rosin, sardine sandwich, blue cheese, pickled cabbage** NOTE DETAILS

Why do Miss Tourees and Miss Russet not want to share Wolfie's lunch with him? **The food has a bad smell.** DRAW CONCLUSIONS

Think and Respond

Have students answer the questions on the inside back cover of the book.

1. **The magnifying glass, hat, and overcoat that the cat is wearing are clues that this is a mystery.**
2. **Possible response: It looks like the word *rectangle*, so I think it means "shaped like a rectangle."**
3. **She sorted the cases into groups. Then she sniffed the cases in one group to find the one that smelled like Wolfie's lunch.**
4. **Possible response: I learned that violins come in child and adult sizes.**
5. **At the beginning of the story, Miss Tourees thinks Wolfie is a talented violinist, but by the end of the story she thinks Wolfie is foolish for not having a tag on his instrument case.**
6. **Possible response: I think the way she solved the mystery was good. I would have handled it the same way.**

SOCIAL STUDIES

Social Studies Connection Have students write about an instrument they play or would like to learn to play. Have them explain why they like the instrument.

School-Home Connection Have students make a list of things that good detectives do.

CM 20 **Independent Practice:** Comprehension Practice page 20

Write

Writer's Craft: Voice

Write a Personal Narrative Have students read page 12 of *The Case of the Missing Case.* Say, *After reading these sentences, I feel that Miss Tourees is irritated by Wolfie. The writer helps to understand these feelings by using the word* huffed *to describe the way Miss Tourees speaks to Wolfie. This shows that she is irritated.* Remind students that good writers share their feelings about a topic through the use of language. Vivid words and descriptive details help make a writer's voice strong. Writers also use sense words to show how someone or something feels, looks, sounds, or smells. Have students write a personal narrative describing a time they lost something and how they found it. Remind students to use punctuation along with words and phrases to express their feelings.

Lost and Found

Fluency

Oral Reading Fluency

CM 21

Assessment Use Oral Reading Fluency Practice page 21 to assess students' oral reading fluency.

Name ____________________

The Case of the Missing Case
Vocabulary

▶ **Match the words with their meanings.**

1. assistant	A. certainly
2. case	B. sure
3. definitely	C. helper
4. returned	D. investigation
5. specific	E. investigator
6. positive	F. given back
7. detective	G. exact

▶ **Imagine that you are a detective. What tools would you need to solve cases? Draw a picture of the tools you would use to solve a mystery and make sure to describe them using at least two vocabulary words.**

Caption: **Possible response: To begin solving a case, I would need a computer to find specific facts.**

▲ Vocabulary Practice p. 19

Name ____________________

The Case of the Missing Case
Comprehension

▶ **Answer the questions below.**

1. How did Wolfie's case get into the storeroom?
   The janitor took it out of Wolfie's dressing room and put it there.
2. What is a storeroom? What smaller words make up the word *storeroom*?
   A storeroom is a room in the basement of a house used to store things; store and room.
3. How does Miss Tourees sort the violin cases?
   by size, color, shape, label, and smell
4. What kind of character do you think Wolfie is, based on how he acts in the story?
   Possible response: Wolfie is annoying. He also seems rude on page 8 when he says he is "not a child."

▶ **Write your answer to the question below.**

How does Miss Tourees find Wolfie's case?
Possible response: She finds out that Wolfie has his smelly lunch in his case; she and Miss Russet sniff until they find the case.

▲ Comprehension Practice p. 20

Name ______________________________

Match the words with their meanings.

1. assistant
2. case
3. definitely
4. returned
5. specific
6. positive
7. detective

A. certainly
B. sure
C. helper
D. investigation
E. investigator
F. given back
G. exact

Imagine that you are a detective. What tools would you need to solve cases? Draw a picture of the tools you would use to solve a mystery and make sure to describe them using at least two vocabulary words.

Caption: ______________________________

____

____

Name ______________________________

Answer the questions below.

1. How did Wolfie's case get into the storeroom?

____

____

2. What is a storeroom? What smaller words make up the word *storeroom*?

____

____

3. How does Miss Tourees sort the violin cases?

____

____

4. What kind of character do you think Wolfie is, based on how he acts in the story?

____

____

Write your answer to the question below.

How does Miss Tourees find Wolfie's case?

____

____

____

Name ______________________________

"Could you be more specific?" asked Miss Tourees. She grabbed her 11
detective notebook to write down clues. 17

"My violin case is missing from the symphony hall," said Wolfie. 28

"Was your violin in the case?" asked Miss Tourees. 37

"No, thank goodness. It's priceless," said Wolfie. "The case is not worth 49
much. I don't think it would be of any special interest to a thief. Still, it's 65
important to me that the case be returned." 73

"I'm sure I can solve the case of the missing case," said Miss Tourees. 87
"I'll meet you at the hall in ten minutes!" 96

Miss Tourees met Wolfie onstage exactly ten minutes later. He was there 108
with his assistant, Miss Russet. 113

"Let's get right to work," said the detective. "Where did you see the case 127
last?" 128

"Mr. B handed me the case during the rehearsal for Fiddle Fest 500," 141
said Miss Russet. "I'm sure I put it in his dressing room." 153

"Are you positive?" asked Miss Tourees. 159

"Yes, I'm positive. I definitely put the case in the dressing room," replied 172
Miss Russet. 174

FLUENCY SCORE

Total Words Read Per Minute ________

Number of Errors – ________

Number of Words Read Correctly ________

Basement Basketball

by Kay Livorse

SUMMARY After being cooped up inside all winter, Maddie and her brother Mark are suffering from cabin fever. Then their aunt has a great idea: to create a basketball court in the basement!

- **Focus Skill:** Narrative Elements
- **Focus Strategy:** Summarize
- **Vocabulary:** professional, familiar, captain, pretended, aimed, monitor

Differentiated Instruction

Independent Reading Have students read the story independently. Encourage students to write down questions to ask if they do not understand a part of the story.

Prepare to Read

Preview/Set Purpose Have students examine the cover and title as well as the illustrations and chapter headings in the book. Point out that this is a realistic fiction book, which tells a made-up story about something that could really happen. Have students predict what the book will be about.

Narrative Elements

Teach/Model Display a story map like the one below. Tell students that good readers think about narrative elements such as characters, setting, and plot to help them understand what they read. Explain that characters are the people the story is about; setting is the time and place in which the story happens; and plot is the events that happen in a story. Tell students that a plot often includes a problem and a solution and has a beginning, middle, and an end. Read pages 2–4 of *Basement Basketball* with students. Say, *This chapter begins with Mark and Maddie suffering from cabin fever. Aunt Lin suggests that they play basketball inside.* Fill in the beginning and middle parts of the plot as shown.

Beginning	→	Middle	→	End
Mark and Maddie have cabin fever.	→	Aunt Lin suggests that they play basketball.	→	**Mark and Maddie learn about basketball and stop feeling grumpy.**

Practice/Apply Have students read pages 5–6 and identify the end of the plot in the chapter.

Summarize

Teach/Model Tell students that good readers pause every few pages to summarize important story events. Explain that summarizing is retelling the most important parts of a story in your own words. Have students read page 2. Say, *After I read page 2, I list the most important points: Aunt Lin notices that Maddie and Mark are grumpy because they have not been able to go out and play.* Point out that summarizing can also help readers understand narrative elements of a story.

Practice/Apply Have students read and summarize pages 4–5.

Build Vocabulary

Help students use context to understand key vocabulary. Have them answer each question by reading the words and sentences that surround the word.

professional (p. 7)	Are you a *professional* basketball player if you play in the park with your brother? **No, professionals play for a career.**
familiar (p. 15)	If something is *familiar*, does that mean you have never seen it before? **No, something familiar is something you know well.**
captain (not in text)	What does the *captain* of a team do? **leads the team**
pretended (p. 2)	How would you act if you *pretended* to play basketball? **You would act as if you were playing, but would not really be playing.**
aimed (p. 15)	If the basketball star *aimed* before taking the shot, what would he be doing? **trying to get the shot in**
monitor (not in text)	Why would a parent *monitor* everyone in the playroom? **to keep everyone safe**

CM 25 **Independent Practice:** Vocabulary Practice page 25

Read and Comprehend

Monitor Comprehension

Guide students as they read the text, using the following questions:

Pages

2–3 Describe Aunt Lin. **She is good at thinking of ideas.** NARRATIVE ELEMENTS Focus Skill

4–6 Why did Dr. Naismith invent basketball? **His students were tired of being indoors.** MAIN IDEA AND DETAILS

How did Dr. Naismith's students solve the problem of having to climb to get the ball? **They cut the bottom out of the basket.** PROBLEM-SOLUTION

7–10 How does Aunt Lin hang the basket? **She ties it to a post.** IMPORTANT DETAILS

11–13 How does Aunt Lin use tape? **She puts lines on the floor for the court.** NOTE DETAILS

14–16 Why does the ball start to go into the basket more often? **The kids have practiced shooting it.** DRAW CONCLUSIONS

Think and Respond

Have students answer the questions on the inside back cover of the book.

1. **She tells Mark and Maddie about Dr. Naismith because his students had a problem similar to their problem.**
2. **The characters cannot go outside to play. They solve their problem by playing basketball in the basement.**
3. **Cabin fever is a phrase used for getting very tired of staying inside.**
4. **You know that they spend a lot of time with her because she knows how the children usually act.**
5. **The children's game was the same because they used a ball and a basket. It was different because their basket was a laundry basket, not a peach basket.**
6. **Possible response: I would like to play basement basketball with them because it seems like a fun game.**

MATH **Math Connection** Have students list all the games they know. Have them use a Venn diagram to sort their lists into inside games, outside games, and both.

School-Home Connection Have students ask a family member what inside games he or she knows.

CM 26 **Independent Practice:** Comprehension Practice page 26

Write

Writer's Craft: Voice

Write a Story Have students read page 2 of *Basement Basketball.* Say, *The writer uses descriptive words like* restless, ignore, annoyed, grumpy, *and* tired *to show how Mark and Maddie are feeling.* Explain that good writers express their voice through vivid details and descriptive language. Have students write a story that takes place on a rainy day. Encourage them to use a story map like the one below to help them organize their thoughts. Remind students to use vivid, descriptive words to express their voice.

Beginning → **Middle** → **Ending**

Fluency

Oral Reading Fluency

CM 27

Assessment Use Oral Reading Fluency Practice page 27 to assess students' oral reading fluency.

Name ______________________

Basement Basketball
Vocabulary

▶ **Write the word that best completes each sentence.**

professional familiar captain pretended aimed monitor

1. The basket looks **familiar** to Mom because she uses it often.
2. Jon **pretended** he didn't hear his mother calling.
3. The coach will **monitor** our game closely.
4. Sarah leads her team in the game, just like a **captain**.
5. People watch **professional** basketball games at Madison Square Garden.
6. The All-Star **aimed** for the basket and scored.

▶ **Write a paragraph about a time you played a sport or game. Use at least three vocabulary words.**

Possible response: Last year I had a soccer game where I had to take a big shot. It was the end of the game, and it was just me and the goalkeeper. I aimed and shot the ball—and I missed! Our team lost! The captain told me that people who play professional soccer miss shots all the time. I felt a little better then.

▲ Vocabulary Practice p. 25

Name ______________________

Basement Basketball
Comprehension

▶ **Imagine you are Mark or Maddie and your parents have come home. Summarize your day for them.**

Possible response: At first we were grumpy because we were tired of being inside, but Aunt Lin said we should invent a new game to play. She told us how people began to play basketball. We found our own basket and used tape and balls to play in the basement. We made up some new rules. It was fun!

▶ **Complete a story map for *Basement Basketball.***

Setting	Characters
inside the house on a rainy day	Aunt Lin, Mark, Maddie, Mom, Dad

Beginning: Problem
Mark and Maddie are bored and grumpy.

Middle: Main Events/Details
Aunt Lin tells Mark and Maddie how basketball was invented.

Ending: Solution to Problem
The characters figure out how to play basketball in their basement.

▲ Comprehension Practice p. 26

Name ______________________________

▶ **Write the word that best completes each sentence.**

professional familiar captain pretended aimed monitor

1. The basket looks ____________________ to Mom because she uses it often.
2. Jon ____________________ he didn't hear his mother calling.
3. The coach will ____________________ our game closely.
4. Sarah leads her team in the game, just like a ____________________.
5. People watch ____________________ basketball games at Madison Square Garden.
6. The All-Star ____________________ for the basket and scored.

▶ **Write a paragraph about a time you played a sport or game. Use at least three vocabulary words.**

____

____

____

____

____

____

Name ______________________________

▶ **Imagine you are Mark or Maddie and your parents have come home. Summarize your day for them.**

____

____

____

____

____

▶ **Complete a story map for *Basement Basketball*.**

Setting

Characters
Aunt Lin, Mark, Maddie, Mom, Dad

Beginning:
Problem
Mark and Maddie are bored and grumpy.

Middle:
Main Events/Details

Ending:
Solution to Problem

Name ______________________________

Aunt Lin watched as the children played. She said that 10
they looked just like professional players. Soon they got better, 20
and the ball went in the basket more often. It was a good game. 34
"Thanks, Aunt Lin," said Mark. 39

"You're the best aunt ever," said Maddie. "You helped us 49
invent a new game—almost." 54

Soon Mom and Dad came home. Both children spoke so 64
fast that Mom laughed. "Stop, stop! You'd better show us," 74
she said. 76

Maddie and Mark were very excited. They showed their 85
parents basement basketball. They talked as they took turns 94
shooting. Maddie aimed and shot. She made the basket. Their 104
parents applauded. 106

"I thought that basket looked familiar! I'm glad I didn't 116
throw it out," said Mom. 121

"Dad," said Mark, "could we have a backboard?" 129

"I think we could, Mark. You could use a big, heavy 140
piece of cardboard. It would keep the basket still. You can 151
practice this winter. In the spring, we can shoot together 161
outside," said Dad. 164

FLUENCY SCORE

Total Words Read Per Minute ________

Number of Errors − ________

Number of Words Read Correctly ________

The Derby

by Max Winter

SUMMARY Danny and his father work together to build a soapbox car out of scraps or metal, being careful to follow the All-American Soapbox Derby rules for size and construction. Danny loses the Derby to another boy with a faster car, but he learns a valuable lesson about sportsmanship.

- **Focus Skill:** Elements of Nonfiction
- **Focus Strategy:** Adjust Reading Rate
- **Vocabulary:** ancient, host, stadium, compete, earned, record, ceremonies, medals

Differentiated Instruction

Paired Reading Organize students in pairs. Have them read the selection aloud to each other, alternating with each page. After students have finished reading, have them discuss a part of the story they found interesting.

Prepare to Read

Preview/Set Purpose Have students look at the cover and illustrations in the book and predict what the story is about. Tell students that the story takes place in the past, when a sport called soapbox racing was popular.

Elements of Nonfiction

Focus Skill

Teach/Model Explain that a book or article that gives facts about real people and events is nonfiction. Information in nonfiction texts usually is divided into sections with headings. Other elements of nonfiction include photographs and the captions below them. Tell students that, although *The Derby* is a realistic fiction story, it contains elements of nonfiction. Read page 4 with students. Note that the first paragraph is fiction, but the diagram looks like something from a nonfiction book. Set these in a chart side by side, as shown below.

Elements of Nonfiction	Elements of Fiction
diagram of soapbox derby racer	Narrator writes to derby officials.

Practice/Apply Have partners look for other elements of nonfiction as they read the story.

Adjust Reading Rate

Focus Strategy

Teach/Model Remind students that good readers adjust their reading rate when they come across difficult passages that have new facts or difficult words. Tell students that reading these passages slowly will help them understand important ideas. Have them read page 11. Say, *This page has some unfamiliar dates and numbers. I am going to read these two paragraphs slowly so that I understand what they say.*

Practice/Apply Tell partners to choose a passage they feel may be difficult. Have them alternate reading the passage aloud, slowly and clearly.

Build Vocabulary

Help students use context to understand key vocabulary. Have them answer each question by reading the words and sentences that surround the word.

ancient (p. 6)	Why might an *ancient* coin be dirty? **because it is very old**
host (p. 14)	What does the *host* of an event provide? **the place**
stadium (p. 13)	Why might you go to a *stadium*? **to see a sporting event**
compete (p. 11)	What is being tested when a person decides to *compete*? **a skill**
earned (p. 15)	Are people proud if they have *earned* something? **Yes, because they have gained something**
record (p. 9)	Why might you want to set a *record*? **to be the best in something**
ceremonies (p. 15)	What honors are sometimes given out at *ceremonies*? **awards**
medals (p. 15)	What material are *medals* made of? **metal**

CM 31 **Independent Practice:** Vocabulary Practice page 31

Read and Comprehend

Monitor Comprehension

Guide students as they read the text, using the following questions:

Pages

2–3 Why could Danny compete at the All-American Soapbox Derby? **He won a local race.** CAUSE-EFFECT

4–6 What two materials must be used to make a soapbox derby racer? **wood and rubber** USE GRAPHIC AIDS

How did Danny make his car for less than $10? **He used materials from a junkyard.** PROBLEM-SOLUTION

7–10 Why did Danny think the soapbox derby record was slipping away? **He had problems in his first test drive.** DRAW CONCLUSIONS

11–13 Name two elements of nonfiction on page 11. **the date, 1933, and the facts about the history of the soapbox derby** ELEMENTS OF NONFICTION (Focus Skill)

How do you know that Danny was confident of winning? **He said his car was the best, and he didn't think he needed to practice.** CHARACTER'S EMOTIONS

14–16 Why do you think the winner spoke with Danny? **Possible response: because he liked racing with Danny** EXPRESS PERSONAL OPINION

Think and Respond

Have students answer the questions on the inside back cover of the book.

1. **Possible response: I was surprised that he did not win because he was ahead of the other cars until the end.**
2. **He followed the plan that the derby officials sent him.**
3. **This story is mainly about a boy's dream of winning the soapbox derby.**
4. **Possible response: It teaches you to be a good sport whether you win or lose.**
5. **They were both made out of crates.**
6. **Possible response: I like to participate in these things. It is exciting to compete, whether you win or lose.**

ART **Art Connection** Tell students to list traits they would want their dream derby car to have. Tell them to think about the color and design of the car.

School-Home Connection Have students share *The Derby* with a family member and talk about how Danny and his dad worked together to build the car.

CM 32 **Independent Practice:** Comprehension Practice page 32

Write

Writer's Craft: Voice

Write a Personal Narrative Have students reread page 16 of *The Derby.* Note the sentence, "I turned around, and what do you know?" and the phrase, "got to know each other a little." Explain that these express an informal, friendly voice. Display a voice web like the one below. Tell students that good writers express voice, or feeling, with the words and punctuation they choose. Have students write a personal narrative about a time they took part in a competition. Explain that a personal narrative is about something that a writer has experienced. Encourage students to use a voice web like the one below to help them get started.

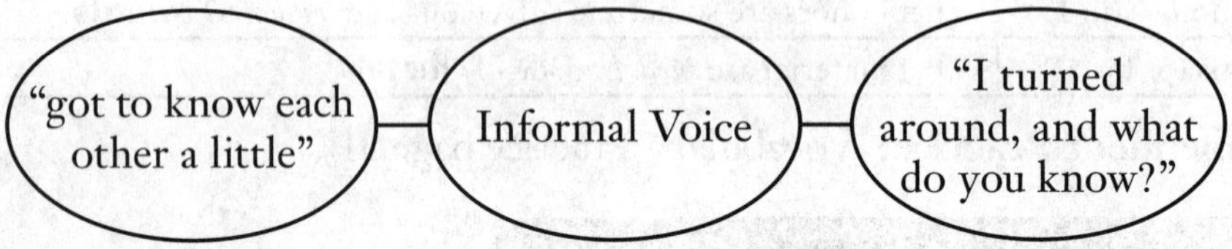

Fluency

Oral Reading Fluency

CM 33

Assessment Use Oral Reading Fluency Practice page 33 to assess students' oral reading fluency.

Name ____________

The Derby
Vocabulary

▶ **Write the word that best describes each list of words.**

ancient	host	stadium	compete
earned	record	ceremonies	medals

1. money, good grades, first prize: things that are **earned**
2. fossils, dinosaurs, cavemen: things that are **ancient**
3. gold, silver, bronze: types of **medals**
4. race, talent show, baseball game: places to **compete**
5. wedding, awards, graduation: types of **ceremonies**
6. seats, field, giant screen: things in a **stadium**
7. party, game show, Olympics: events with a **host**
8. highest jump, fastest sprint, longest throw: examples of a **record**

▶ **Describe what might happen at a race, using three words from the box.**

Possible response: The big race was held in the stadium. The runners went to the line; they were ready to compete. The crowd stood and waited for the start. Maybe someone would set a new record!

Name ____________

The Derby
Comprehension

▶ **Underline the incorrect part in each sentence and rewrite it correctly.**

1. Danny saw his first soapbox derby race on August 15, 1937, <u>in Columbus, Ohio.</u>
in Akron, Ohio
2. Danny's <u>brother</u> helped him get ready for the derby.
father
3. <u>Danny</u> used his tools to make his car.
Danny's father
4. A Dayton newspaper <u>reporter</u> had the idea for the derby.
photographer
5. Danny's dad was <u>angry with</u> him after the race.
proud of

▶ **Look at the diagram on page 4 of *The Derby.* Describe what a soapbox derby car should look like.**

Possible response: The car should be made of wood, with rubber wheels. It should have a steering wheel. It can't be more than 30 inches high, and it has to be at least 30 inches wide. It can't be more than 72 inches long.

▲ Vocabulary Practice p. 31

▲ Comprehension Practice p. 32

Name ______________________________

▶ **Write the word that best describes each list of words.**

ancient	**host**	**stadium**	**compete**
earned	**record**	**ceremonies**	**medals**

1. money, good grades, first prize: things that are

   ____

2. fossils, dinosaurs, cavemen: things that are ____________________

3. gold, silver, bronze: types of ____________________

4. race, talent show, baseball game: places to ____________________

5. wedding, awards, graduation: types of ____________________

6. seats, field, giant screen: things in a ____________________

7. party, game show, Olympics: events with a ____________________

8. highest jump, fastest sprint, longest throw: examples of a

   ____

▶ **Describe what might happen at a race, using three words from the box.**

____

____

____

____

Name ______________________________

▶ Underline the incorrect part in each sentence and rewrite it correctly.

1. Danny saw his first soapbox derby race on August 15, 1937, in Columbus, Ohio.

____

2. Danny's brother helped him get ready for the derby.

____

3. Danny used his tools to make his car.

____

4. A Dayton newspaper reporter had the idea for the derby.

____

5. Danny's dad was angry with him after the race.

____

▶ Look at the diagram on page 4 of *The Derby*. Describe what a soapbox derby car should look like.

____

____

____

____

Name ____________________

Dad got a crazy smile on his face. He was pointing at something. I 14
looked and saw a pile of junk. Then I saw what he saw. He walked over 30
and lifted two boxes into the air. One was an old canned peas crate. The 45
other was a box for grape soda. I knew it, then and there. Those boxes 60
would make my soapbox racer. Sure, the boxes were dirty. They also 72
looked ancient. But I knew they would work. 80

The next day, I started building the car. I measured the crates to 93
make sure they were OK. One of the crates was too tall. I measured 42 108
inches with Dad's tape measure, and then I made a mark. I drew a line 123
across the crate. This would tell Dad where to saw the wood. Dad 136
wouldn't let me use his tools. He and Mom were sure I would hurt 150
myself. I would hold the boards in place. Dad would hammer nails 162
into them. 164

FLUENCY SCORE

Total Words Read Per Minute ____________

Number of Errors − ____________

Number of Words Read Correctly ____________

Turtle at Home

by Rachel Kranz

SUMMARY When Rosa discovers a small turtle while picnicking in the woods with her aunt, she wants to take the turtle home with her. After her aunt explains that taking the turtle may seriously affect the turtle population, Rosa decides to let the turtle go.

- **Focus Skill:** Author's Purpose
- **Focus Strategy:** Reread to Clarify
- **Vocabulary:** message, litter, trained, patiently, wise, eager

Differentiated Instruction

Paired Reading Organize students in pairs. Have them read the selection aloud to each other, alternating with each page. After students have finished reading, have them discuss in pairs a part of the story they found interesting.

Prepare to Read

Preview/Set Purpose Invite students to share what they know about turtles. Then guide them in previewing the book's title, cover, chapter headings, and pictures. Ask students what they think they will learn from this book.

Focus Skill

Author's Purpose

Teach/Model Tell students that good readers try to figure out the author's purpose in what they read. Explain to students that authors have different reasons or purposes for writing, such as to entertain, to inform, and to persuade. Tell them that authors sometimes combine more than one purpose in a single story. Read pages 5–6 of *Turtle at Home.* Say, *On these two pages, the author gives us facts about turtles: Baby turtles' shells are still soft; female turtles' eyes are brown–the males have red eyes. By telling us these facts, the author is trying to inform us about box turtles. So, the author's purpose on these pages is to inform the readers.* Work with students to fill in the chart below.

Author's Purpose: to inform readers about box turtles		
Detail	**Detail**	**Detail**
Baby turtles' shells are still soft.	Female turtles' eyes are brown.	Males have red eyes.

Practice/Apply Have partners add to the chart as they read.

Focus Strategy

Reread to Clarify

Teach/Model Tell students that good readers reread to understand important ideas and parts of a story that seem confusing. Read page 2. Say, *When I read page 2, I do not understand what Rosa means when she says, "It's not that." To clear up my confusion, I reread page 2. When I do, I see that Rosa's aunt thinks that Rosa is eager because she is hungry. I see that Rosa says, "It's not that" to explain to her aunt that she is not eager to eat but is eager to walk around because she has never been in the woods before.* Point out that rereading to clarify can also help readers identify the author's purpose in a selection.

Practice/Apply Have partners stop whenever they read something confusing. Tell them to reread to clear up their confusion.

Build Vocabulary

Help students use context to understand key vocabulary. Have them answer each question by reading the words and sentences that surround the word.

message (p. 14)	When you give someone a *message*, what do you give them? **information**
litter (p. 16)	What might be left as *litter*? **Possible response: paper plates, napkins, plastic bags**
trained (p. 6)	How does being *trained* to do a job help you? **It helps you know the job better and do it well.**
patiently (not found in book)	Why might a hungry person have a hard time waiting *patiently* for dinner? **because the person has to wait for food without complaining**
wise (p. 9)	Is a *wise* idea a good idea? **yes**
eager (p. 2)	When are you *eager* about something? **when you want to do it**

CM 37 **Independent Practice:** Vocabulary Practice page 37

Read and Comprehend

Monitor Comprehension

Guide students as they read the text, using the following questions:

Pages

2–3 Why did Rosa and her aunt go to the woods? **for a picnic** STORY EVENTS

4–7 How can Aunt Lydia tell that the turtle is an adult? **It has a hard shell.** RECALL DETAILS
Which page would you reread to find out if a turtle is male or female? **page 6** REREAD TO CLARIFY (Focus Strategy)

8–11 List three things that a turtle eats. **Possible response: lettuce, mushrooms, worms** CLASSIFY

12–13 In a turtle's lifetime, how many of its babies usually survive? **two or three** RECALL DETAILS
How did the number of turtles in the woods change between Aunt Lydia's childhood and now? **There were many more turtles during Aunt Lydia's childhood.** COMPARE-CONTRAST

14–16 Why does the author have Rosa leave the turtle in the woods? **to protect it and keep it in its natural home** AUTHOR'S PURPOSE (Focus Skill)

Think and Respond

Have students answer the questions on the inside back cover of the book.

1. **This story takes place in the woods.**
2. **Possible response: The author wants to teach us that turtles need to be in the wild to live a long and happy life.**
3. **It is important to leave the turtle in the woods because it will be happier there and will lay eggs to make more turtles.**
4. **Possible response: Aunt Lydia probably does not like the idea of selling turtles in pet stores because they should be left in the wild.**
5. **Rosa wants to take the turtle home in the beginning, but in the end she decides to leave it in the woods.**
6. **Possible response: I would feel happy to leave the turtle in its home.**

Science Connection Have students make a list of facts about turtles.

School-Home Connection Have students brainstorm with a friend or relative at home ways to help protect endangered species.

CM 38 **Independent Practice:** Comprehension Practice page 38

Write

Writer's Craft: Effective Paragraphs

Write Directions Have students read the "What Rosa Does" section of *Turtle at Home*. Say, *The writer explains what has happened to turtles over the years, and why it is important to leave them in their natural habitat. Aunt Lydia explains important details, or gives directions, about what must be done to protect turtles.* Remind students that a group of sentences that explain sequence and give details about how to do something create an effective paragraph. Have students think about planning a picnic for their friends. Have them write a paragraph giving directions about what to pack for the picnic and how to choose a location for the picnic.

Planning a Picninc . . .

Fluency

Oral Reading Fluency

CM 39

Assessment Use Oral Reading Fluency Practice page 39 to assess students' oral reading fluency.

Name ______________________

Turtle at Home
Vocabulary

▶ **Write the word that best completes the sentence.**

message litter trained patiently wise eager

1. We were careful not to leave **litter** from the picnic on the ground.
2. Before Joan began to talk, she waited **patiently** for José to finish speaking.
3. Sometimes the author's purpose is a **message** that the author wants to give the reader.
4. Tyler was **eager** to take his new puppy home.
5. When Rachel heard her grandmother's stories, she understood that her grandmother was very **wise**.
6. Because Monica's father had been **trained** as a scientist, he knew a lot about stars.

▶ **Write a paragraph about the importance of keeping the environment healthy. Use at least two vocabulary words.**

Possible response: One of the most important and simple things we can do to protect the environment is not to litter. If everyone in the world understood this basic message we would have a clean and safe place to live for many years.

Name ______________________

Turtle at Home
Comprehension

▶ **Tell whether each statement is True or False. If the statements are false, rewrite them to make them true.**

1. Aunt Lydia knows so many facts about turtles because she is trained as a scientist.
   True
2. At the end of the story, Rosa decides to take the turtle home as a pet.
   False; she decides to leave the turtle where it is.
3. The author may have written this story in order to teach her readers about turtles and to persuade them that turtles should be protected.
   True
4. At the end of the story, Aunt Lydia tells Rosa that she should never go back to visit the turtle.
   False; Aunt Lydia tells her she can go back and visit.

▶ **List reasons Rosa decides not to take the turtle home.**

Possible responses: The turtle needs special food; If she moved the turtle, it would try to get home; The turtle would not be able to lay eggs, so baby turtles would not be born.

▲ Vocabulary Practice p. 37

▲ Comprehension Practice p. 38

Name ______________________________

▶ **Write the word that best completes the sentence.**

message	litter	trained	patiently	wise	eager

1. We were careful not to leave ______________ from the picnic on the ground.

2. Before Joan began to talk, she waited ______________ for José to finish speaking.

3. Sometimes the author's purpose is a ______________ that the author wants to give the reader.

4. Tyler was ______________ to take his new puppy home.

5. When Rachel heard her grandmother's stories, she understood that her grandmother was very ______________.

6. Because Monica's father had been ______________ as a scientist, he knew a lot about stars.

▶ **Write a paragraph about the importance of keeping the environment healthy. Use at least two vocabulary words.**

____

____

____

____

Name ______________________________

▶ **Tell whether each statement is True or False. If the statements are false, rewrite them to make them true.**

1. Aunt Lydia knows so many facts about turtles because she is trained as a scientist.

____

2. At the end of the story, Rosa decides to take the turtle home as a pet.

____

3. The author may have written this story in order to teach her readers about turtles and to persuade them that turtles should be protected.

____

4. At the end of the story, Aunt Lydia tells Rosa that she should never go back to visit the turtle.

____

▶ **List reasons Rosa decides not to take the turtle home.**

____

____

____

Name ______________________________

"If you take the turtle, she might not lay any more eggs. Soon there 14
won't be many turtles left in the forest," Aunt Lydia explained. 25

"Right now, a few missing turtles may not matter," she said. "It 37
will make a big difference later on. When I was little, there were 50
many turtles here. Now there are very few." 58

"I understand, Aunt Lydia," Rosa said. She lifted the turtle and 69
looked into its eyes. It was like the turtle was sending her a message. 83
Rosa lowered her hand to the ground. 90

"Good-bye," she said sadly. "I hope you can still find your home." 102

"I think she can," Aunt Lydia said. She put her arm around Rosa. 115
"You hardly moved her." 119

Aunt Lydia and Rosa watched the little turtle. Rosa saw how the 131
sun lit up the turtle's back. 137

"I'm proud of you, Rosita," Aunt Lydia said. "I think you did the 150
right thing." 152

Rosa looked up at Aunt Lydia. "I know I did the right thing, Aunt 166
Lydia," she said. 169

FLUENCY SCORE

Total Words Read Per Minute ____________

Number of Errors – ____________

Number of Words Read Correctly ____________

Better Than Gold

by Carolyn Clark

Realistic Fiction

SUMMARY In 1925, the Walters family moves from California to Nome, Alaska, for a summer in search of gold. Mr. Walters does not find gold, but he and his family find something better than gold—they gain friends among the Native Alaskans and other residents. Their neighbors end up helping them build a home so they can remain in Alaska.

- **Focus Skill:** Word Relationships
- **Focus Strategy:** Use Decoding/Phonics
- **Vocabulary:** telegraph, drifts (n.), temperature, guided, trail, splinters

Differentiated Instruction

Paired Reading Organize students in pairs. Have them read the story aloud to each other, alternating with each page. After children have finished reading, have them discuss in pairs a part of the story they found interesting.

Prepare to Read

Preview/Set Purpose Have students preview the book by looking at the pictures and have them guess where the story takes place. Ask them what they think they will learn from this book.

Word Relationships

Focus Skill

Teach/Model Display a word relationship chart similar to the one below. Tell students that good readers pay attention to the way words are related to help them understand what they read. Explain that words can be synonyms, antonyms, or homophones. Synonyms are words that mean the same, e.g., *father* and *dad*. Antonyms are words that have opposite meanings, e.g., *many* and *few*. Homophones are words that sound the same but have different spellings and meanings, e.g., *there* and *their*.

Word Relationships	Word From the Text	Related Word
synonyms	father	dad
antonyms	many	few
homophones	there	their

Practice/Apply Have partners read the chart above and find a synonym for *job*, an antonym for *heavy*, and a homophone for *right*.

Use Decoding/Phonics

Focus Strategy

Teach/Model Tell students that good readers use different methods to figure out words they don't know. Point out that they can use letter sounds, or phonics, to help them figure out new words. Model using *farther*. Say, *To figure out what this word is, I'll think about the sounds the letters stand for. The first syllable could be /fär/ and I know that* th *says* /th/. *So I can blend the letters together to form* farth. *I also know that* -er *says* /er/. *When I blend the sounds together it says* farther.

Practice/Apply Have partners use the letter sounds to learn new words. Have them explain how they figured out the words.

Build Vocabulary

Help students use context to understand key vocabulary. Have them answer each question by reading the words and sentences that surround the word.

telegraph (p. 5)	Why might you go to the *telegraph* office? **to send information**
drifts (p. 16)	What is the weather like if you are caught in *drifts*? **cold and snowing**
temperature (p. 9)	Describe *temperature*. **It is how hot or cold something is.**
guided (p. 10)	How is someone *guided*? **Someone who knows the way leads.**
trail (p. 10)	What word means the same as *trail*? **path**
splinters (p. 15)	What material do you get lots of *splinters* from? **wood**

CM 43 **Independent Practice:** Vocabulary Practice page 43

Read and Comprehend

Monitor Comprehension

Guide students as they read the text, using the following questions:

Pages

2–3 Why did the Walters family move to California? **to find gold** CHARACTER'S MOTIVATIONS Summarize page 3 in two sentences. **In 1925, Mr. Walters read in the newspaper about a 1898 Gold Rush in Nome, Alaska. He learned that a team of sled dogs carried medicine to Nome through snow and helped save people's lives.** SUMMARIZE

4–6 If you travel from Nome to San Francisco, in what direction are you going? **south** USE GRAPHIC AIDS

7–9 How would you describe Mr. Wolfe? **Possible response: Mr. Wolfe seems kind and generous.** EXPRESS PERSONAL OPINION

10–12 What did Mr. Walters do in Nome that he also did in San Francisco? **looked for gold** COMPARE-CONTRAST

13–16 What is a synonym for *gold*? **treasure** WORD RELATIONSHIPS Focus Skill

Think and Respond

Have students answer the questions on the inside back cover of the book.

1. **This story takes place in the year 1925, in California and Alaska.**
2. **The first time the word *game* is used on page 9 it means something you play. The second time it is used it means wild animals that are hunted.**
3. **The family moves to Alaska in search of gold and finds that friends are better than gold.**
4. **Possible response: I learned that Alaska is cold and the people there are friendly.**
5. **Possible response: Houses in California and Alaska would be alike because they would both have walls and a roof. They would be different because the houses in Alaska would have thicker walls to keep out the cold.**
6. **Possible response: I would choose to make friends because we would help each other and have fun together.**

Social Studies Connection Have students write five questions about Alaska.

School-Home Connection Ask students to have a discussion with someone at home about how people are different.

CM 44 **Independent Practice:** Comprehension Practice page 44

Write

Writer's Craft: Effective Paragraphs

Write a Paragraph of Information Have students read page 16 of *Better Than Gold.* Say, *In the last paragraph, the author writes about Mr. Walters losing interest in his search for gold. The paragraph begins with the sentence "Daddy lost most of his interest in searching for gold after we got the house." This is the topic sentence and tells us what the paragraph will be about. The next few sentences, known as detail sentences, give us information on why Mr. Walters lost interest in gold. Mr. Walters learns that gold alone would not make his life better. He realizes that his true wealth is his family.* Have students read the first paragraph on page 15 of *Better Than Gold.* Have them figure out the topic of the paragraph, and write the topic sentence and details of this paragraph in the diagram below.

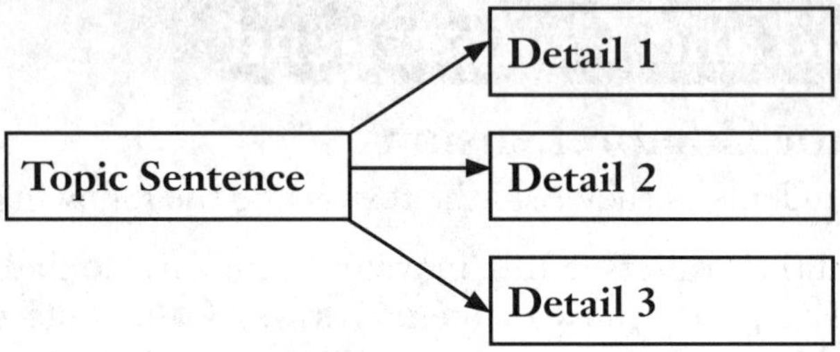

Fluency

Oral Reading Fluency

CM 45

Assessment Use Oral Reading Fluency Practice page 45 to assess students' oral reading fluency.

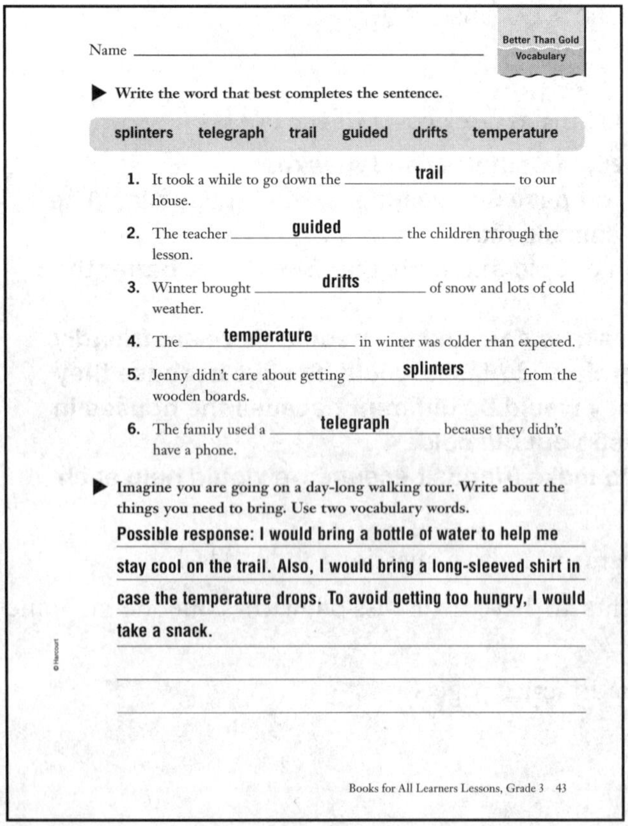

Name ______________________

Better Than Gold
Vocabulary

▶ **Write the word that best completes the sentence.**

splinters telegraph trail guided drifts temperature

1. It took a while to go down the **trail** to our house.
2. The teacher **guided** the children through the lesson.
3. Winter brought **drifts** of snow and lots of cold weather.
4. The **temperature** in winter was colder than expected.
5. Jenny didn't care about getting **splinters** from the wooden boards.
6. The family used a **telegraph** because they didn't have a phone.

▶ **Imagine you are going on a day-long walking tour. Write about the things you need to bring. Use two vocabulary words.**

Possible response: I would bring a bottle of water to help me stay cool on the trail. Also, I would bring a long-sleeved shirt in case the temperature drops. To avoid getting too hungry, I would take a snack.

© Harcourt

Books for All Learners Lessons, Grade 3 43

▲ Vocabulary Practice p. 43

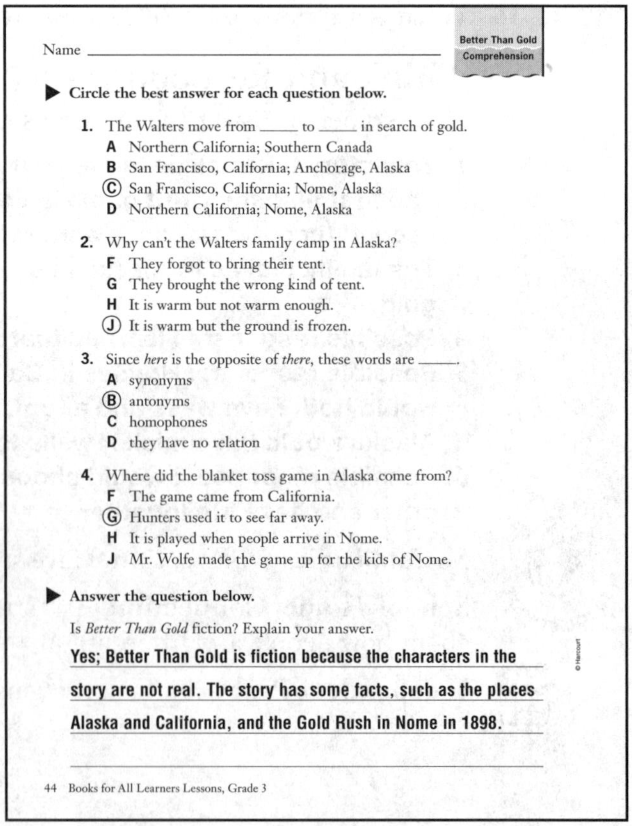

Name ______________________

Better Than Gold
Comprehension

▶ **Circle the best answer for each question below.**

1. The Walters move from ____ to ____ in search of gold.
   - A Northern California; Southern Canada
   - B San Francisco, California; Anchorage, Alaska
   - (C) San Francisco, California; Nome, Alaska
   - D Northern California; Nome, Alaska
2. Why can't the Walters family camp in Alaska?
   - F They forgot to bring their tent.
   - G They brought the wrong kind of tent.
   - H It is warm but not warm enough.
   - (J) It is warm but the ground is frozen.
3. Since *here* is the opposite of *there*, these words are ____.
   - A synonyms
   - (B) antonyms
   - C homophones
   - D they have no relation
4. Where did the blanket toss game in Alaska come from?
   - F The game came from California.
   - (G) Hunters used it to see far away.
   - H It is played when people arrive in Nome.
   - J Mr. Wolfe made the game up for the kids of Nome.

▶ **Answer the question below.**

Is *Better Than Gold* fiction? Explain your answer.

Yes; Better Than Gold is fiction because the characters in the story are not real. The story has some facts, such as the places Alaska and California, and the Gold Rush in Nome in 1898.

© Harcourt

44 Books for All Learners Lessons, Grade 3

▲ Comprehension Practice p. 44

Name ______________________________

▶ **Write the word that best completes the sentence.**

splinters telegraph trail guided drifts temperature

1. It took a while to go down the ____________ to our house.
2. The teacher ____________ the children through the lesson.
3. Winter brought ____________ of snow and lots of cold weather.
4. The ____________ in winter was colder than expected.
5. Jenny didn't care about getting ____________ from the wooden boards.
6. The family used a ____________ because they didn't have a phone.

▶ **Imagine you are going on a day-long walking tour. Write about the things you need to bring. Use two vocabulary words.**

____

____

____

____

____

____

Name ______________________________

▶ **Circle the best answer for each question below.**

1. The Walters move from _____ to _____ in search of gold.
- **A** Northern California; Southern Canada
- **B** San Francisco, California; Anchorage, Alaska
- **C** San Francisco, California; Nome, Alaska
- **D** Northern California; Nome, Alaska

2. Why can't the Walters family camp in Alaska?
- **F** They forgot to bring their tent.
- **G** They brought the wrong kind of tent.
- **H** It is warm but not warm enough.
- **J** It is warm but the ground is frozen.

3. Since *here* is the opposite of *there*, these words are _____.
- **A** synonyms
- **B** antonyms
- **C** homophones
- **D** they have no relation

4. Where did the blanket toss game in Alaska come from?
- **F** The game came from California.
- **G** Hunters used it to see far away.
- **H** It is played when people arrive in Nome.
- **J** Mr. Wolfe made the game up for the kids of Nome.

▶ **Answer the question below.**

Is *Better Than Gold* fiction? Explain your answer.

____

____

____

____

Name ____

Daddy went to the telegraph office to see about making travel 11
plans. He came home whistling a cheery tune. “We’re all set!” he 23
said. “We leave next week.” 28

The next week, we packed our things and drove to San 39
Francisco. When we got there, Daddy sold our Model T Ford, and 51
we boarded a ship headed for Nome, Alaska. 59

Ships were pretty much the only way to get people and cargo to 72
Nome. There weren’t many roads. There weren’t any railroads or 82
passenger airplanes, either. 85

Julia and I had a great time on the ship. It was mostly meant to 100
carry cargo, so there were lots of big boxes that were good for 113
playing hide-and-seek. 115

Mama gave us geography lessons. She showed us a map of 126
where we were going. From San Francisco, we headed north on the 138
Pacific Ocean. We passed Oregon, Washington, and some of 147
Canada. 148

The farther north we went, the colder it got. It was May, but we 162
were shivering. 164

FLUENCY SCORE

Total Words Read Per Minute ____________

Number of Errors − ____________

Number of Words Read Correctly ____________

Realistic Fiction

My Pet Project

by Debra Hershkowitz

SUMMARY When Teddy receives a new camera for his birthday, he decides to make a special project of photographing friends' pets. This book includes snapshots and descriptions of all the animals he photographed.

- **Focus Skill:** Author's Purpose
- **Focus Strategy:** Use Text Structure and Format
- **Vocabulary:** creature, curious, marine (adj.), delicate, collapsed, survived

Differentiated Instruction

Paired Reading Organize students in pairs. Have them read the selection aloud to each other, alternating with each page. After partners have finished reading, have them discuss a part of the story they found interesting.

Prepare to Read

Preview/Set Purpose Have students examine the cover and title. Have them preview the book by looking at the headings and photographs. Then have students explain how these text features give them clues about the story. Point out that this is a realistic fiction book.

Focus Skill

Author's Purpose

Teach/Model Tell students that authors have different reasons, or purposes, for writing. Authors may write to entertain, inform, persuade, or give directions. Tell them that authors sometimes combine more than one purpose in a single story. Have students read the second paragraph on page 3. Say, *In this paragraph the author gives us information about how a camera works. The author's purpose in telling us these facts is to inform us about cameras.*

Practice/Apply After partners have read *My Pet Project*, have them discuss ways the author writes to inform.

Focus Strategy

Use Text Structure and Format

Teach/Model Explain to students that paying attention to the format, or text features, of a selection will help them locate important information. Tell students that text features they should look for include headings, subheadings, photographs, captions, and graphic aids. Point out that when good readers come across a part of a story they find confusing, they can try to figure out how that part of the story connects to the previous part. Have students read pages 2–4. Say, *When I read page 4, I do not understand why they are looking at baby pictures. Then I remember that on page 2 Teddy got a new camera. On page 3, Teddy's aunt shows him how to use it. That helps me to understand that on page 4, Teddy is trying to decide what makes a good photograph, so he looks at pictures that others have taken.*

Practice/Apply Have partners stop whenever they find something confusing and use this strategy to locate and understand information.

Build Vocabulary

Help students use context to understand key vocabulary. Have them answer each question by reading the words and sentences that surround the word.

creature (p. 11)	What kind of *creature* barks and pants? **dogs**
curious (p. 7)	If you are *curious* about rabbits, how do you find out more about them? **by asking questions, reading**
marine (p. 11)	Name a *marine* animal. **Possible response: fish**
delicate (p. 2)	Which is more *delicate*: a rock, a vase, or a Frisbee? **a vase**
collapsed (p. 4)	If I ran home and *collapsed* on my bed, how was I feeling? **tired**
survived (p. 15)	I planted three flowers and two of them are still living. How many *survived*? **two**

CM 49

Independent Practice: Vocabulary Practice page 49

Read and Comprehend

Monitor Comprehension

Guide students as they read the text, using the following questions:

<u>Pages</u>

2–5 What is Aunt Dee's job? **She is a photographer.** NOTE DETAILS

What did Aunt Dee get Teddy for his birthday? **a camera** STORY EVENTS

6–7 How did Teddy get Sadie to sit still? **He threw the Frisbee to her until she got tired.** PROBLEM-SOLUTION

What is Astor's main character trait? **He is curious.** CHARACTERIZATION

8–9 Why didn't Teddy take a picture of the parrots in their cage? **He couldn't get a clear picture of the parrots without including the bars of the cage.** ANALYZE

10–11 Why did Jane name her guinea pig "Spike"? **because his fur sticks out and looks like spikes** CHARACTER'S MOTIVATIONS

What is the author's purpose in including the story about Spike whistling? **to entertain** AUTHOR'S PURPOSE (Focus Skill)

12–16 Where did Sami and Alex take their animals to win a blue ribbon? **the county fair** SETTING

Think and Respond

Have students answer the questions on the inside back cover of the book.

1. **The "I" in the story is Teddy.**
2. **Possible response: I think she told so much about Teddy's photography strategies because it helps the reader learn about taking photos.**
3. **Possible response: It gives the reader a hint that the story will be about pets.**
4. **Possible response: I learned that taking pictures is a lot of work but fun.**
5. **They were alike because they were always moving. They were different because the fish are in a tank and the dog is not.**
6. **Possible response: They look at magazines and photo albums to see what makes a good picture.**

SCIENCE

Science Connection Have students use the Internet to research a famous photographer.

School-Home Connection Have students work with someone at home to find old photographs and compare how they are different from new photographs.

CM 50

Independent Practice: Comprehension Practice page 50

Write

Writer's Craft: Effective Paragraphs

Write a Summary Tell students that an effective paragraph is organized with a topic sentence, followed by details that support the main idea. Explain that a summary is one type of effective paragraph. A summary includes all the important points from the original text. Have students reread page 7 of *My Pet Project.* Say, *If I were writing a summary, I would include these important points: Will's cat, Astor, is curious; sometimes he is hard to find; Teddy tried to photograph him being curious.* Have students write a summary of pages 14–15 of *My Pet Project.*

Fluency

Oral Reading Fluency

CM 51

Assessment Use Oral Reading Fluency Practice page 51 to assess students' oral reading fluency.

Name ______________________

My Pet Project
Vocabulary

▶ **Write the word that best completes each sentence.**

creature curious marine delicate collapsed survived

1. Lou's friends were **curious** to see the pictures he took at his birthday party.
2. After the storm, Paco was glad to see that the carrots in his backyard had **survived**.
3. I went diving in the ocean and saw many **marine** animals, such as eels and rays.
4. You have to be careful when handling some animals because they can be very **delicate**.
5. A blue whale is the largest **creature** on Earth.
6. I laughed so hard at my brother that I almost **collapsed**.

▶ **Think about your favorite animal. Write about that animal using at least two vocabulary words.**

Possible response: Monkeys are very cute creatures. They are curious and like to climb trees.

▲ Vocabulary Practice p. 49

Name ______________________

My Pet Project
Comprehension

▶ **Use the section headings in *My Pet Project* to help fill in the blanks.**

The author wrote this story for two reasons: to **inform** readers about cameras and animals, and to **entertain** them. At the begining of the story, Teddy gets a camera from his aunt, who is a **photographer**. He decides to take pictures of his friends' pets.

First, Teddy tries to take a picture of **Joey**'s dog, Sadie. Teddy throws her a **Frisbee** to make sure she sits still. Then Teddy takes a picture of Will's **cat**, Molly's **parrots**, and Paco's **rabbit**. When he goes to visit Jane and her guinea pig, he laughs at Spike because **he kind of whistles**. Two other animals discussed in this section are a **dog** and a **cat**. At the end of the story, Teddy gives **a copy of each pet's photograph** to each of his friends.

▶ **Answer the question below.**

Why do you think the author includes the diagram on page 3?

It helps the author to inform readers about cameras.

▲ Comprehension Practice p. 50

Name ______________________________

▶ **Write the word that best completes each sentence.**

creature **curious** **marine** **delicate** **collapsed** **survived**

1. Lou's friends were ______________ to see the pictures he took at his birthday party.
2. After the storm, Paco was glad to see that the carrots in his backyard had ______________.
3. I went diving in the ocean and saw many ______________ animals, such as eels and rays.
4. You have to be careful when handling some animals because they can be very ______________.
5. A blue whale is the largest ______________ on Earth.
6. I laughed so hard at my brother that I almost ______________.

▶ **Think about your favorite animal. Write about that animal using at least two vocabulary words.**

____

____

____

____

Name ______________________________

▶ **Use the section headings in *My Pet Project* to help fill in the blanks.**

The author wrote this story for two reasons: to ______________ readers about cameras and animals, and to ______________ them. At the begining of the story, Teddy gets a camera from his aunt, who is a ______________. He decides to take pictures of his friends' pets.

First, Teddy tries to take a picture of ______________'s dog, Sadie. Teddy throws her a ______________ to make sure she sits still. Then Teddy takes a picture of Will's ______________, Molly's ______________, and Paco's ______________. When he goes to visit Jane and her guinea pig, he laughs at Spike because ______________. Two other animals discussed in this section are a ______________ and a ______________. At the end of the story, Teddy gives ______________ to each of his friends.

▶ **Answer the question below.**

Why do you think the author includes the diagram on page 3?

____

____

Name ______________________________

My camera lesson began. Aunt Dee showed me what each 10
part of a camera does. 15

Sometimes too much light hits the film. That makes the picture too 27
light. Sometimes there's too little light. That makes the picture too 38
dark. Aunt Dee set my camera for the light in the room. After that, she 53
put film in the camera. 58

Then it was my turn. Aunt Dee showed me how to hold the 71
camera. I practiced looking through the viewfinder. 78

We talked about what makes good photos. First, we looked at old 90
family pictures. Boy, she had gotten some funny shots of me! We 102
nearly collapsed from laughing so hard. Other pictures helped me 112
remember times that I had forgotten. 118

We also looked at magazines. We saw pictures of people, places, 129
and animals. Aunt Dee showed me what made each picture a good 141
photo. 142

Then she said, "Look at things around you. Take pictures of 153
anything you like. Most of all, have fun!" 161

FLUENCY SCORE

Total Words Read Per Minute ____________

Number of Errors – ____________

Number of Words Read Correctly ____________

Dinosaur Times

by Vita Jimenez

SUMMARY Dinosaurs large and small roamed the earth in Mesozoic times, then mysteriously disappeared. Maps and illustrations help convey what scientists know about dinosaurs and how they study them.

- **Focus Skill:** Word Relationships
- **Focus Strategy:** Self-Question
- **Vocabulary:** brunch, peaceful, omelet, erupting, lava, escape

Differentiated Instruction

Paired Reading Organize students into pairs. Have them read the selection aloud to each other, alternating with each page. After students have finished reading, have them talk about the parts of the story that they found most interesting.

Prepare to Read

Preview/Set Purpose Have students examine the cover and title. Have students preview the book by looking at the headings, illustrations, and maps. Point out that this is a nonfiction book, which is a book about real people, places, and events. Ask them what they think they will learn from this book.

Word Relationships

Focus Skill

Teach/Model Display a word relationship chart similar to the one below. Tell students that good readers pay attention to the way words are related to help them understand what they read. Explain that some words are related because they are synonyms or have almost the same meaning. Also point out that other words are related because they are antonyms or have opposite meanings. Ask students to scan page 6 to find synonyms. Then have them scan page 10 to find antonyms. Put the words they find in the word relationship chart.

Word Relationships	Description	Page	Word from text	Related word
synonyms	Two words with the same meaning	6	**remains**	**traces**
antonyms	Two words with opposite meanings	10	**slow**	**quick**

Practice/Apply Have students add synonyms and antonyms from the book to the word relationship chart as they read.

Self-Question

Focus Strategy

Teach/Model Tell students that when good readers come across a part of a story they find confusing, they ask themselves questions to try to clear up the confusion. Read page 3 and say, *When I read page 3, I ask myself: What did the author mean when she said it might have been more peaceful when dinosaurs lived? I know that people are not peaceful when they fight, so the author must mean that dinosaurs did not fight. That may be why the land was more peaceful.*

Practice/Apply Have partners choose a page to read. Then have them use self-questioning to clarify something they find confusing.

Build Vocabulary

Help students use context to understand key vocabulary. Have them answer each question by reading the words and sentences that surround the word.

brunch (p. 13)	How can you tell that *brunch* does not combine "breakfast" and "dinner"? **No part of the word "dinner" is used in "brunch," but part of "lunch" is used.**
peaceful (p. 3)	What word means the opposite of *peaceful*? **noisy, stormy**
omelet (p. 13)	Why would someone think that an *omelet* could be made with dinosaur eggs? **Omelets are made with eggs.**
erupting (p. 14)	What word or phrase can you use in place of *erupting*? **blowing out, exploding**
lava (p. 14)	What color is *lava* when it erupts? **orange or red like fire**
escape (p. 14)	How do we know that dinosaurs did not *escape* death? **They are extinct.**

CM 55 **Independent Practice:** Vocabulary Practice page 55

Read and Comprehend

Monitor Comprehension

Guide students as they read the text, using the following questions:

Pages

2–5 How were dinosaurs able to roam all over the earth? **Land on Earth was joined together.** DRAW CONCLUSIONS

6–7 List three examples of fossils from the Age of Dinosaurs. **bones, footprints, and leaf prints in stone** CLASSIFY/CATEGORIZE

Why are dinosaur fossils important to scientists? **They help us understand the shape and size of dinosaurs.** INTERPRETING INFORMATION

8–9 What did scientists learn about the eating habits of dinosaurs? **More dinosaurs ate meat than ate plants.** IMPORTANT DETAILS

10–13 What antonyms are used on page 10 to describe the different types of dinosaurs? **big and slow; small and quick** WORD RELATIONSHIPS Focus Skill

What does the name *Tyrannosaurus rex* mean? **"all powerful king"** MAIN IDEA AND DETAILS

14–16 What theory do scientists have about why dinosaurs disappeared? **Dust blocked the sun's rays so that plants died, and there was nothing to eat.** CAUSE-EFFECT

Think and Respond

Have students answer the questions on the inside back cover of the book.

1. **It is divided into six parts. Each part has a question and an answer about a dinosaur topic.**
2. ***Dirt* is a synonym for *soil* or *ground*.**
3. **We have learned about dinosaurs by studying fossils.**
4. **Possible response: I learned how scientists find fossils.**
5. **The Brachiosaurus was similar to the Tyrannosaurus rex because they were both big dinosaurs. They were different because the Brachiosaurus ate plants while the Tyrannosaurus rex ate meat.**
6. **Possible response: I would give the fossil to a museum.**

SCIENCE

Science Connection Ask students to create a list of three things they would like to know about dinosaurs. Then have students research dinosaurs to find the information.

School-Home Connection Have students make a list of natural history museums in their area. Have them write a letter to someone in their family to ask if they will take them to one of the museums.

CM 56 **Independent Practice:** Comprehension Practice page 56

Write

Writer's Craft: Effective Paragraphs

Write a How-To Essay Have students read pages 8–9 of *Dinosaur Times.* Point out how the writer gives information on how to find fossils. Say, *This is an example of a how-to paragraph. It begins with the topic sentence, "To find fossils, scientists go to places where they think fossils might be buried." This is followed by details on how scientists go about looking for fossils.* Point out the use of the words *next* and *finally.* Explain that these time-order words help make sentences flow smoothly from one to the next. Then have students write a paragraph on how to find the best book about dinosaurs. To help them get started, encourage them to put their ideas in a graphic organizer like the one below.

Topic sentence → Detail 1 → Detail 2 → Detail 3

Fluency

Oral Reading Fluency

CM 57

Assessment Use Oral Reading Fluency Practice page 57 to assess students' oral reading fluency.

Name ______________________ Dinosaur Times Vocabulary

▶ **Write the word that best completes each sentence.**

brunch omelet peaceful erupting lava escape

1. Many people believe that dust from an active, **erupting** volcano might have caused dinosaurs to become extinct.
2. I planned to **escape** from summer camp because it was not fun.
3. Tom used a dozen eggs to make an **omelet**.
4. The hot **lava** flowing down the sides of the volcano burned everything in its path.
5. It was too late to eat breakfast, so I ate a big **brunch** instead.
6. An antonym for loud or angry is **peaceful**.

▶ **Draw an illustration of a Brachiosaurus. Then write a caption that uses at least two vocabulary words.**

Caption: **Possible response: Brachiosaurus was a peaceful dinosaur that might have liked eating leaves for brunch.**

Books for All Learners Lessons, Grade 3 55

▲ Vocabulary Practice p. 55

Name ______________________ Dinosaur Times Comprehension

▶ **Circle the best answer for the questions below.**

1. Which theory best explains why dinosaurs became extinct?
   - A A meteor killed them.
   - B Man began to kill the dinosaurs.
   - (C) Dust of some kind blocked the sun's rays.
   - D An earthquake destroyed the land.
2. Why should scientists handle fossils carefully?
   - F They might have dangerous germs.
   - G They don't want to get cut by the jagged edges.
   - H They are usually too large to carry.
   - (J) They are so old they might fall apart.
3. Which group of words below contains antonyms for *escape*?
   - A leave, run away, depart
   - (B) capture, remain, return
   - C break out, leave, capture
   - D remain, return, break away
4. What is the main idea of this story?
   - F Dinosaurs lived a long time ago and were mean animals.
   - (G) Dinosaurs roamed the earth and mysteriously disappeared.
   - H Scientists use dinosaur bones to solve crimes today.
   - J People believe dinosaurs can still be found on Earth.

▶ **Explain how asking yourself questions as you read helped you to understand *Dinosaur Times.***

Possible response: I did not understand why people did not know how dinosaurs became extinct, so I asked myself why they wouldn't know. I know that dinosaurs are not alive today, and scientists have to do research to know what happened to them. That means that scientists probably don't have enough information to prove how dinosaurs became extinct.

56 Books for All Learners Lessons, Grade 3

▲ Comprehension Practice p. 56

Name ________________________________

▶ **Write the word that best completes each sentence.**

brunch **omelet** **peaceful** **erupting** **lava** **escape**

1. Many people believe that dust from an active, ______________ volcano might have caused dinosaurs to become extinct.
2. I planned to ______________ from summer camp because it was not fun.
3. Tom used a dozen eggs to make an ______________.
4. The hot ______________ flowing down the sides of the volcano burned everything in its path.
5. It was too late to eat breakfast, so I ate a big ______________ instead.
6. An antonym for loud or angry is ______________.

▶ **Draw an illustration of a Brachiosaurus. Then write a caption that uses at least two vocabulary words.**

Caption: ________________________________

____

____

Name ______________________________

▶ **Circle the best answer for the questions below.**

1. Which theory best explains why dinosaurs became extinct?
A A meteor killed them.
B Man began to kill the dinosaurs.
C Dust of some kind blocked the sun's rays.
D An earthquake destroyed the land.

2. Why should scientists handle fossils carefully?
F They might have dangerous germs.
G They don't want to get cut by the jagged edges.
H They are usually too large to carry.
J They are so old they might fall apart.

3. Which group of words below contains antonyms for *escape*?
A leave, run away, depart
B capture, remain, return
C break out, leave, capture
D remain, return, break away

4. What is the main idea of this story?
F Dinosaurs lived a long time ago and were mean animals.
G Dinosaurs roamed the earth and mysteriously disappeared.
H Scientists use dinosaur bones to solve crimes today.
J People believe dinosaurs can still be found on Earth.

▶ **Explain how asking yourself questions as you read helped you to understand *Dinosaur Times*.**

____

____

____

Name ______________________________

Scientists have slowly been able to figure out how Earth **10**
must have looked back then. For example, it might have been more **22**
peaceful. Scientists have been able to figure this out using signs **33**
that have been left behind on Earth. **40**

There are many ideas about what Earth used to look like. Many **52**
scientists think that long, long ago one part of Earth was all land. **65**
The other part was all water. Dinosaurs lived all over the land. **77**

Over time, the land broke up into big parts. These parts became **89**
what we now call *continents.* The water parts became the different **100**
oceans, rivers, and other bodies of water that we know today. **111**

Dinosaurs still lived on the land after it broke apart. When they **123**
died, they left bones behind. They also left other signs that they had **136**
lived there. Scientists have found dinosaur bones in almost every **146**
country on Earth. This means that there are very few places where **158**
dinosaurs did not roam. **162**

FLUENCY SCORE

Total Words Read Per Minute ____________

Number of Errors — ____________

Number of Words Read Correctly ____________

Fiction

Deefer

by Kay Livorse

SUMMARY Carrie wishes for a puppy. Her mother and grandmother like dogs but recall the difficulties of raising a puppy. A visit to an animal shelter brings a compromise—an older dog—and the need to work out ways to share in the care and retraining of their new pet.

- **Focus Skill:** Decode Long Words
- **Focus Strategy:** Use Context to Confirm Meaning
- **Vocabulary:** firm, confident, comfortable, approach, equipment, program, appointment

Differentiated Instruction

Paired Reading Organize students in pairs. Have them read the selection aloud to each other, alternating with each page. After students have finished reading, have them discuss in pairs a part of the story they found interesting.

Prepare to Read

Preview/Set Purpose Have students examine the cover and title. Then have students look at the pictures in the book. Remind students that this is a fiction book, which is a book about made-up people and situations. Ask students what they think the book might be about.

Focus Skill

Decode Long Words

Teach/Model Display a chart similar to the one below. Tell students that good readers decode long words by breaking them into smaller parts. Some long words are made up of two different words, as in the word *backgrounds*, which is made up of *back* and *grounds*. Other words are made up of a root word with an ending added on, as in the word *suggested*, which is made up of the root word *suggest* with *-ed* added on.

Word from Text	Word Parts	Word Meaning
backgrounds	back/grounds	**all of a thing's experience and origins**
suggested	suggest/ed	**mentioned for thinking about**

Practice/Apply Have partners decode the following words from *Deefer*: *crawled, responsible, adoption, identification, surroundings.*

Focus Strategy

Use Context to Confirm Meaning

Teach/Model Tell students that good readers use context to make sure they know the meaning of a long word. Explain that context is the meanings of other words in the sentence. Have students read page 6. Point out that the word *suggested* is one long word in this paragraph. Say, *When I read a long word, I break it into parts. I can break* suggested *into* suggest *and* ed. *I can see that* suggest *is the root word. It means to propose for thinking about, as in to* suggest *a tasty kind of food or a good book. Now I look at the context, or the other words in the sentence around* suggested *to make sure I've found the correct meaning. These are the words around* suggested: *"Why don't we think about an older dog?"* suggested *Mom. I can tell from this sentence that* suggested *means to bring up a new idea to think about. Now I substitute my meaning for* suggested, *which is* proposed, *into the sentence. "Why don't we think about an older dog?"* proposed *Mom. From the context, I can tell that my meaning of* suggested *is correct.*

Practice/Apply Have partners use context to confirm the meaning of *expected* on page 14.

Build Vocabulary

Help students use context to understand key vocabulary. Have them answer each question by reading the words and sentences that surround the word.

firm (p. 16)	What word would you use to describe a *firm* person? **Possible response: steady**
confident (p. 5)	What words describe a person who is the opposite of *confident*? **nervous, unsure**
comfortable (p. 15)	What does *comfortable* mean? **feeling safe**
approach (p. 4)	Should you *approach* an angry dog? **No.**
equipment (p. 10)	Name types of *equipment* that you need to take care of a dog. **food bowl, water bowl, leash, identification tag**
program (p. 15)	Why would you put a dog on a special *program*? **to train the dog in a certain way**
appointment (p. 10)	What does *appointment* mean? **to set or decide a time and date**

CM 61 **Independent Practice:** Vocabulary Practice page 61

Read and Comprehend

Monitor Comprehension

Guide students as they read the text, using the following questions:

Pages

2–4 What is a good place to go to get a dog? **an animal shelter** MAIN IDEA AND DETAILS
Name one good thing and one bad thing about having a puppy. **Possible response: They are cute animals. They are a lot of work.** CLASSIFY-CATEGORIZE

5–7 Why do some people give their dogs to shelters? **They have to move, they change their minds, the dog might be bad with kids, or the dog might be too big.** RECALL DETAILS

8–10 How are adoption counselors helpful? **They help people choose the right dog for their families.** CHARACTERIZATION

11–13 Why does Grandma say that the dog has bad manners? **The dog begs for food at the table.** DRAW CONCLUSIONS

14–16 Why does Mom think the dog is misbehaving? **because the dog eats the food in the kitchen and does not know what is expected of it** MAKE INFERENCES
What is the root word in *remembered*? **remember** DECODE LONG WORDS (Focus Skill)

Think and Respond

Have students answer the questions on the inside back cover of the book.

1. **Possible response: The titles tell the reader the main topic of each section.**
2. **Possible response: I broke the word into syllables and blended them together.**
3. **The main idea of this book is that taking care of a dog is a lot of work.**
4. **Possible response: I learned that animal shelters have adoption counselors who help people choose the right dog for their family.**
5. **The dog's life was better because she was lonely at the animal shelter, and now she has a new family to walk her and play with her.**
6. **Possible response: I would adopt an older dog from an animal shelter because there are many that need homes.**

SOCIAL STUDIES

Social Studies Connection Have students research in the school library about different dog breeds and pick the one they like best.

School-Home Connection Have students make a list of things they need to have at home if they have a dog. Have them read their lists to a family member.

CM 62 **Independent Practice:** Comprehension Practice page 62

Write

Writer's Craft: Sentence Fluency

Write a How-To Essay Have students read page 15 of *Deefer.* Say, *The writer tells us how to train a pet. The paragraph begins with a topic sentence which is followed by details on how Carrie and her family plan to train their dog.* Tell students that effective paragraphs are made up of sentences that have varied lengths and that flow from one to the other. Effective paragraphs are also organized in a logical pattern, using sequence words like *first*, *next*, and *finally*. Sequence is very important when writing an essay about how to do something. Have students write a paragraph on how to keep a pet healthy. Remind them to write a topic sentence and use different kinds of sentences in their writing.

Fluency

Oral Reading Fluency

CM 63

Assessment Use Oral Reading Fluency Practice page 63 to assess students' oral reading fluency.

Name ______________________ Deefer Vocabulary

▶ **Match each word with its meaning.**

C 1. firm	A. having trust or faith	
A 2. confident	B. anything made for a particular use	
F 3. comfortable	C. uncompromising and strict	
G 4. approach	D. a plan of what will be done	
B 5. equipment	E. an arrangement to meet or to do something	
D 6. program	F. feeling at ease	
E 7. appointment	G. a way of doing something or to get close	

▶ **Create a poster and slogan for an animal shelter. Use at least two of the vocabulary words in your advertisement.**

Possible response: At our animal shelter, we make you comfortable with your pet. No appointment needed!

▲ Vocabulary Practice p. 61

Name ______________________ Deefer Comprehension

▶ **Circle the best answer for each question below.**

1. What is the best reason to adopt a dog instead of buying a puppy?
   A Adopting a dog is cheaper.
   B Puppies are a lot of work.
   C Adopted dogs are nicer than puppies.
   (D) You can rescue an older, already trained dog.
2. Why didn't Grandma want to get a puppy?
   F Grandma was allergic to dogs.
   (G) Grandma thought she couldn't give a puppy enough attention.
   H Grandma wanted to get a kitten instead.
   J Grandma thought Carrie was not old enough for a puppy.
3. Which groups of words below are synonyms of the word *firm*?
   A angry, mad, upset
   (B) strict, strong, steady
   C tired, slowed, sleepy
   D loud, noisy, roaring
4. What is the purpose of the *program* that the family puts Deefer on?
   F to teach the dog new and exciting tricks
   G to teach the dog how to hunt for its own food
   (H) to teach the dog how to have better manners
   J to teach the dog how to steal food from the table

▶ **Answer the question below.**

Is *Deefer* fiction or nonfiction? Explain your answer.

Possible response: Fiction; the characters and events of the story are not real.

▲ Comprehension Practice p. 62

Name ______________________________

Match each word with its meaning.

____	**1.** firm	**A.**	having trust or faith
____	**2.** confident	**B.**	anything made for a particular use
____	**3.** comfortable	**C.**	uncompromising and strict
____	**4.** approach	**D.**	a plan of what will be done
____	**5.** equipment	**E.**	an arrangement to meet or to do something
____	**6.** program	**F.**	feeling at ease
____	**7.** appointment	**G.**	a way of doing something or to get close

Create a poster and slogan for an animal shelter. Use at least two of the vocabulary words in your advertisement.

____

____

Name ______________________________

▶ **Circle the best answer for each question below.**

1. What is the best reason to adopt a dog instead of buying a puppy?
- **A** Adopting a dog is cheaper.
- **B** Puppies are a lot of work.
- **C** Adopted dogs are nicer than puppies.
- **D** You can rescue an older, already trained dog.

2. Why didn't Grandma want to get a puppy?
- **F** Grandma was allergic to dogs.
- **G** Grandma thought she couldn't give a puppy enough attention.
- **H** Grandma wanted to get a kitten instead.
- **J** Grandma thought Carrie was not old enough for a puppy.

3. Which groups of words below are synonyms of the word *firm*?
- **A** angry, mad, upset
- **B** strict, strong, steady
- **C** tired, slowed, sleepy
- **D** loud, noisy, roaring

4. What is the purpose of the *program* that the family puts Deefer on?
- **F** to teach the dog new and exciting tricks
- **G** to teach the dog how to hunt for its own food
- **H** to teach the dog how to have better manners
- **J** to teach the dog how to steal food from the table

▶ **Answer the question below.**

Is *Deefer* fiction or nonfiction? Explain your answer.

____

____

Name ______________________________

When they got home, Grandma called the vet to ask what 11
kind of food to get for the new dog. 20

Carrie and her mother went shopping for the dog. They bought 31
food, bowls, a leash, a collar, and some treats. Carrie could hardly 43
wait to get back to her dog. 50

While they were having lunch, their new dog begged for food at 62
the table. 64

"This won't do," said Grandma. "Our dog must have good 74
manners just like we do, Carrie. She'll have to wait in the kitchen 87
while we finish lunch." 91

After their meal, Carrie brought the lunch dishes into the 101
kitchen. "Oh, no!" she thought. The dog had eaten the last of the 114
pumpkin pie from the kitchen table. 120

Grandma came in. "Where is the pie?" she asked. "Where is the 132
rest of the turkey? Is anything else missing?" 140

"Don't be angry, Grandma," begged Carrie. 146

"Grandma isn't angry," said Carrie's mom. "She is 154
remembering what living with a dog is like." 162

FLUENCY SCORE

Total Words Read Per Minute ____________

Number of Errors — ____________

Number of Words Read Correctly ____________

The Way Home

by Meish Goldish

SUMMARY Ted, a nine-year old boy, and his older brother, Mike, go hiking on a family camping trip and aren't sure how to get back to camp until Ted makes a compass.

- **Focus Skill:** Sequence
- **Focus Strategy:** Read Ahead
- **Vocabulary:** cartwheel, seriously, mustache, fastened, beyond, collection

Differentiated Instruction

Paired Reading Organize students in pairs. Have them read the story aloud to each other, alternating with each page. After students have finished reading, have them discuss in pairs a part of the story they found interesting.

Prepare to Read

Preview/Set Purpose Have students preview the book by examining the cover and title. Point out that this book is realistic fiction, in which the characters, setting, and plot all could be real, but are not. Have students use the information from the cover to predict what will happen in the story.

Focus Skill

Sequence

Teach/Model Display a chart similar to the one below. Tell students that the sequence of events in a story is the order in which things happen. Sequence is especially important when a writer gives instructions explaining how to do something. Say, *If I wanted to explain how to make a peanut butter and jelly sandwich, the first step would be to gather all of the materials. Next, you would spread the peanut butter on one slice of bread. Third, spread the jelly on the other slice of bread. Then, put the slices of bread together so that the peanut butter and jelly touch. Finally, you would eat the sandwich.* Have partners read page 4 of *The Way Home* and use their own words to fill in the steps for putting up a tent.

Step 1	**Lay the tent out on a flat piece of land.**
Step 2	**Hammer stakes into the ground.**
Step 3	**Attach the tent to the stakes with some rope.**

Practice/Apply Ask partners to read page 4 and fill in the steps to set up camp on a sequence chart.

Focus Strategy

Read Ahead

Teach/Model Tell students that good readers read ahead to help them answer questions that come up as they read. Point out that when good readers become confused, they read the next paragraph or two to figure it out. Have students read page 5. Say, *When I begin reading page 5, I am confused about why Ted wants to go hiking so badly. As I read ahead, I see that Ted likes to collect stones and shells. That explains why Ted wants to go hiking.*

Practice/Apply Have partners stop and use the Read Ahead strategy whenever they find something confusing.

Build Vocabulary

Help students use context to understand key vocabulary. Have them answer each question by reading the words and sentences that surround the word.

beyond (p. 10)	What word can you use in place of *beyond*? **Possible response: past**
cartwheel (p. 2)	Why would you do a *cartwheel*? **Possible response: You are happy.**
collection (p. 5)	What kind of collection do you have? **Possible response: a baseball card collection**
fastened (p. 4)	What word means the opposite of *fastened*? **Possible response: unbuttoned, opened**
mustache (p. 16)	Why might a man get crumbs in his *mustache*? **A mustache is near the mouth.**
seriously (p. 11)	What word means the opposite of *seriously*? **jokingly**

CM 67 **Independent Practice:** Vocabulary Practice page 67

Read and Comprehend

Monitor Comprehension

Guide students as they read the text, using the following questions:

Pages

2–5 Where does this story take place? **at South Lake Camp Grounds** SETTING

What does the family do after Dad and Mike set up the tent? **They eat.** SEQUENCE (Focus Skill)

6–7 Why was Ted looking for stones? **He collects them.** USING PRIOR KNOWLEDGE

Why did Ted want to show his rock to Mrs. Smith? **She would know if it was a fossil.** CHARACTER'S MOTIVATIONS

8–11 According to the picture on page 10, the boys are walking in which direction? **south** USE GRAPHIC AIDS

12–13 Why didn't Ted and Mike wait for the North Star to come out? **They had promised their parents they would come back before dark.** MAKE INFERENCES

14–16 How did Mike and Ted feel about finding their way back to camp? **proud** CHARACTERS' EMOTIONS

Think and Respond

Have students answer the questions on the inside back cover of the book.

1. **The text is divided into three chapters.**
2. **Ted knew that the campsite was north and the lake was south. So he made a compass and walked north.**
3. **Mike learned how magnets and compasses are alike.**
4. **Possible response: I learned that magnets have two poles.**
5. **His first way of finding north was to use the North Star, but the second way was using a magnet.**
6. **Possible response: I would yell for help.**

SCIENCE

Science Connection Collect classroom items, including some that attract magnets. Have students predict which items will attract magnets.

School-Home Connection Have students write directions from your classroom to another place in the school. Have them share the directions with a family member.

CM 68 **Independent Practice:** Comprehension Practice page 68

Write

Writer's Craft: Development

Write a Persuasive Paragraph Have students read page 7 of the selection. Say, *This is an example of a persuasive paragraph. Ted's opinion is that the stone might be a fossil. He gives reasons to support that opinion, such as that the stone might contain a leaf print.* Tell students that a well-developed persuasive paragraph states a clear opinion in the first sentence and supports that opinion with good reasons. Have students list some activities they enjoy. Have them choose one activity to develop and write a paragraph convincing their readers to try it.

Opinion: **Possible response: Listening to music is the best way to spend your free time.**
Reason 1: **Possible response: Music is very relaxing.**
Reason 2: **Possible response: There are many different types to choose from.**
Reason 3: **Possible response: It is available for free on the radio.**

Fluency

Oral Reading Fluency

CM 69

Assessment Use Oral Reading Fluency Practice page 69 to assess students' oral reading fluency.

Name ______________________

The Way Home
Vocabulary

▶ Write the word that best completes each sentence.

cartwheel seriously mustache fastened beyond collection

1. Before she wiped the milk from her lip, it looked as if Mom had a **mustache**.
2. The lake is just **beyond** that big pine tree on the right.
3. I fell down the first time I tried to do a **cartwheel**.
4. Her **collection** grew because she kept bringing rocks home.
5. It was **seriously** scary to be lost in the woods.
6. We **fastened** our bike helmets before riding.

▶ Write a postcard from Ted to his friends. On the left side, draw a picture of South Lake Camp Grounds. On the right side, write a message using at least two vocabulary words.

Possible response: I did a cartwheel when I found a new stone for my collection.

© Harcourt

Books for All Learners Lessons, Grade 3 67

▲ Vocabulary Practice p. 67

Name ______________________

The Way Home
Comprehension

▶ Tell whether each statement is True or False. If a statement is false, rewrite it to make it true.

1. Ted had a collection of stones and shells at his home.
True
2. Mike and Ted both knew that Ted had found a fossil.
False; Ted thought he had found a fossil, but he was not sure. He planned to ask his teacher, Mrs. Smith, about it.
3. Mike used magnets to make a compass, but it turned out that Ted had one in his pocket the whole time.
False; Ted made a compass out of magnets, and Mike had a compass in his pocket the whole time.

▶ Answer the question below.

How could reading ahead help the reader understand what was happening to Ted and Mike during the story?

Possible response: The reader may not understand why Ted was collecting things. Also, the reader might worry that Ted and Mike will not find the way back to camp. Reading ahead would help clear up this confusion.

© Harcourt

68 Books for All Learners Lessons, Grade 3

▲ Comprehension Practice p. 68

Name ______________________________

Write the word that best completes each sentence.

cartwheel seriously mustache fastened beyond collection

1. Before she wiped the milk from her lip, it looked as if Mom had a ______________.

2. The lake is just ______________ that big pine tree on the right.

3. I fell down the first time I tried to do a ______________.

4. Her ______________ grew because she kept bringing rocks home.

5. It was ______________ scary to be lost in the woods.

6. We ______________ our bike helmets before riding.

Write a postcard from Ted to his friends. On the left side, draw a picture of South Lake Camp Grounds. On the right side, write a message using at least two vocabulary words.

	______________ ______________ ______________ ______________ ______________ ______________

Name ______________________________

▶ **Tell whether each statement is True or False. If a statement is false, rewrite it to make it true.**

1. Ted had a collection of stones and shells at his home.

____

____

2. Mike and Ted both knew that Ted had found a fossil.

____

____

3. Mike used magnets to make a compass, but it turned out that Ted had one in his pocket the whole time.

____

____

▶ **Answer the question below.**

How could reading ahead help the reader understand what was happening to Ted and Mike during the story?

____

____

____

____

Name ______________________________

The brothers walked faster. The sun was starting to get low in the sky. **14**
Soon the boys reached the lake. Ted and Mike walked around it. Ted **27**
found some pine cones. He put them in his jacket pocket. **38**
Suddenly, Mike noticed how low the sun was. He looked at his watch. **51**
"Oh, no," he said seriously. "It's late. We said we'd be back before dark. **65**
Let's go." **67**
"Which trail should we take?" asked Ted. "There are so many." **78**
"Well, which one would you pick?" Mike asked. **86**
"Maybe I can figure it out," Ted said. "We walked south to get to the **101**
lake. In order to get back to the campsite, we have to walk north." **115**
"Do you know where north is?" asked Mike. **123**
Ted stopped to think. "If it were dark, we could check the sky and **137**
follow the North Star," he said. **143**
"But it's not dark," said Mike. **149**
"No, but it will be soon," said Ted. **157**
"You're right," said Mike. "We'd better think of something quick!" **167**

FLUENCY SCORE

Total Words Read Per Minute ____________
Number of Errors − ____________
Number of Words Read Correctly ____________

Realistic Fiction

Piano Piece

by Suzanne Martinucci

SUMMARY Adam is reluctant to play the piano at the big concert because he is afraid his stage fright will make him forget the music. When his older brother, Gary, a confident and skilled pianist, sprains his wrist, Adam suggests that the brothers play a duet with one brother using his right hand and the other using the left.

- **Focus Skill:** Prefixes and Suffixes
- **Focus Strategy:** Create Mental Images
- **Vocabulary:** roaming, gym, recite, prefer, perform, enjoying, billions

Differentiated Instruction

Paired Reading Organize students in pairs. Have them read the selection aloud to each other, alternating with each page. After children have finished reading, have them discuss in pairs a part of the story they found interesting.

Prepare to Read

Preview/Set Purpose Have students preview the book by looking at the chapter headings and illustrations. Have them explain how these features will be helpful. Point out that this is realistic fiction, a story about people, places, and events that could be real, but are not. Ask them what they think this book will be about.

Focus Skill

Prefixes and Suffixes

Teach/Model Tell students that you can change the meaning of a word by adding a prefix or a suffix. Explain that prefixes come before a root word while suffixes come after a root word. Point out that learning word parts helps you define new words easily. Display the chart below and tell students that the prefix *un-* means "not." Explain that the suffix *-ous* changes a verb or noun into an adjective, while *-ly* means "how something is done." Work with students to define each word.

Root word	Prefix or Suffix	New Word	Meaning
like	un-	unlike	not like
joy	-ous	joyous	full of joy
quiet	-ly	quietly	done with little sound

Practice/Apply Have students list any words they found in the selection that contain the suffixes *-ly* or *-ous.*

Focus Strategy

Create Mental Images

Teach/Model Tell students that good readers use the Create Mental Images strategy to help them imagine events as they read a story. Point out that good readers stop from time to time and use the author's words to create images in their minds. Have students read page 4. Say, *When I read I notice Gary is talking about a man named Horowitz, but there is no picture of him. I use the description that Gary gives to make a picture of Horowitz in my head. I see a man on stage, shaking as he walks to his piano. There is sweat running down his face, and he is very nervous.*

Practice/Apply Have partners describe the mental images they created from details or events in *Piano Piece.*

Build Vocabulary

Help students use context to understand key vocabulary. Have them answer each question by reading the words and sentences that surround the word.

roaming (p. 3)	Why wouldn't you go *roaming* if you were in a hurry? **It uses up time.**
gym (p. 12)	Explain why someone would go to a *gym*. **Possible response: to play games**
recite (p. 11)	Which of these two things can a person *recite*: a swing set or a speech? **a speech**
prefer (p. 14)	What word or phrase can you use in place of *prefer*? **like better, favor**
perform (p. 3)	Describe what musicians do when they *perform*. **They play music in front of people.**
enjoying (p. 8)	Why would people keep playing a game that they are *enjoying*? **because they were having fun playing it**
billions (p. 6)	What word or phrase means the opposite of *billions*? **few, none**

CM 73 **Independent Practice:** Vocabulary Practice page 73

Read and Comprehend

Monitor Comprehension

After students read the text, have them answer the following questions:

Pages

2–4 What is Adam's problem? **He doesn't want to play at the recital.** MAIN IDEA AND DETAILS

5–7 Why did Adam think that Gary was better than he was? **Gary loved playing for people and never forgot his music.** COMPARE AND CONTRAST

What did Gary suggest that Adam do to solve his problem? **He told Adam to pretend that he was playing at home.** PROBLEM-SOLUTION

8–10 How do you think Adam felt when his father was talking to him? **Possible response: relieved** IDENTIFY WITH CHARACTER

11–13 Where did Adam get the idea to play a duet with Gary? **He saw two kids taking turns playing handball together.** CAUSE-EFFECT

How does the suffix *-ed* change the verb *start* on page 13? **It puts it in the past tense.** PREFIXES AND SUFFIXES (Focus Skill)

14–16 What do you think happened when the boys played their duet? **Possible response: Adam enjoyed himself. They had fun performing together.** MAKE PREDICTIONS

What term on page 16 helps you create a mental image? **Possible response: bowed** CREATE MENTAL IMAGES (Focus Strategy)

Think and Respond

Have students answer the questions on the inside back cover of the book.

1. **The title tells me that the story is about playing the piano.**
2. **The suffix is *-ly*.**
3. **The main idea is that Adam and Gary work as a team to play at the concert.**
4. **Possible response: I learned that it takes a lot of practice to play the piano.**
5. **Possible response: They are alike because they both play the piano well. They are different because Adam is shy and Gary is not.**
6. **Possible response: I am more like Gary because I like to perform.**

Language Arts Connection Have students list the guidelines they would keep in mind if they had to perform for an audience.

School-Home Connection Have students make a list of things they could perform for their family.

CM 74 **Independent Practice:** Comprehension Practice page 74

Write

Writer's Craft: Development

Write a Speech Have students reread page 6. Say, *Gary tells Adam that to solve his stage fright, he should pretend he's home. Then he gives reasons why this works well. This is an example of good development.* Explain that a well-developed speech begins with a statement of the speaker's opinion, followed by reasons and details that support that opinion. Point out that it ends with a request for the audience to do or believe something. Have students imagine that they have to convince a younger friend or sibling to perform at a concert. Have them write a speech using the chart.

Topic:
Reason 1: **Reason 2:** **Reason 3:**

Fluency

Oral Reading Fluency

CM 75

Assessment Use Oral Reading Fluency Practice page 75 to assess students' oral reading fluency.

Name ______________________ Piano Piece Vocabulary

▶ Draw a line to match each word to the related words and phrases.

1. roaming	public show, practice
2. gym	millions, a lot
3. recite	want most, favorite
4. prefer	wandering, strolling
5. perform	play games, whistle blow, basketball net
6. enjoying	memorize, repeat
7. billions	having fun, being happy

▶ Write a paragraph describing some things you might do while roaming around your neighborhood. Use at least three vocabulary words.

Possible response: I stop at the rec center when roaming around my neighborhood. Some people play basketball in the gym, while others recite poetry. My brothers run around enjoying the company of others, but I prefer to be alone.

▲ Vocabulary Practice p. 73

Name ______________________ Piano Piece Comprehension

▶ For each section of *Piano Piece*, list three vivid words or phrases the author used that helped you create a mental image.

Section	Vivid Words
1. Pages 2–4	**Possible response: terrible, roaming, mussed**
2. Pages 5–7	**Possible response: mumbled, deep breath, suggest**
3. Pages 8–10	**Possible response: nervous, hurrying, tripped**
4. Pages 11–13	**Possible response: recite, big weight, listen carefully**
5. Pages 14–16	**Possible response: teased, bowed, nod**

▶ Rewrite the ending of the story. Use vivid words to help your audience create a mental image.

Possible response: After helping Gary, Adam felt confident enough to play his own piece, too. He quickly found the correct ivory keys. It began as soft as a whisper. The audience held its breath to hear every chord. Then the harmony grew until the entire room was bursting with its sweet sound. As his fingers slid across the keys for the final time, the audience clapped more loudly than thunder.

▲ Comprehension Practice p. 74

Name ______________________________

▶ **Draw a line to match each word to the related words and phrases.**

1. roaming	public show, practice
2. gym	millions, a lot
3. recite	want most, favorite
4. prefer	wandering, strolling
5. perform	play games, whistle blow, basketball net
6. enjoying	memorize, repeat
7. billions	having fun, being happy

▶ **Write a paragraph describing some things you might do while roaming around your neighborhood. Use at least three vocabulary words.**

____

____

____

____

____

____

Name ____________________

▶ **For each section of *Piano Piece*, list three vivid words or phrases the author used that helped you create a mental image.**

Section	Vivid Words
1. Pages 2–4	
2. Pages 5–7	
3. Pages 8–10	
4. Pages 11–13	
5. Pages 14–16	

▶ **Rewrite the ending of the story. Use vivid words to help your audience create a mental image.**

Name ______________________________

Mom went back into the house. Gary and Adam played a little 12
basketball. "I heard you practicing yesterday," Gary said. "You know that 23
Copland piece really well." 27

Adam said, "I know I know it. I've played it billions of times. I just get 43
nervous on stage. Even the piano starts to look strange. It's different when I 57
play at home." 60

"Pretend that you *are* home!" Gary said. "That's all I do. Take a deep 74
breath. Have fun with the music." 80

"I can't," Adam mumbled. 84

"Well, I can suggest one other thing," Gary said. "Make sure you know 97
the music cold. Hey look, Dad's home! I hope that means we eat soon!" 111
With that, Gary ran off. 116

Adam followed a little more slowly. "Maybe Gary's right," he thought. "I 128
just need to practice more." 133

Later that night, Adam lay awake in bed. He had called Mr. Peters. He 147
said he would play the Aaron Copland piece. Adam felt very nervous about 160
the upcoming concert, though. 164

FLUENCY SCORE

Total Words Read Per Minute ________

Number of Errors − ________

Number of Words Read Correctly ________

Roberto's Dream

by Rebecca Motil

SUMMARY Roberto Clemente dreamed of becoming a professional baseball player, and he overcame many obstacles to achieve his dream. He played for the Pittsburgh Pirates, leading them to victory in the 1960 World Series. In 1972, he died in a plane crash while bringing supplies to earthquake-stricken Nicaragua.

- **Focus Skill:** Sequence
- **Focus Strategy:** Summarize
- **Vocabulary:** ballhawk, vanish, fault, depend, concentrate, outfielder

Differentiated Instruction

Paired Reading Organize students in pairs. Have them read the book aloud to each other, alternating with each page. After children have finished reading, have them discuss in pairs a part of the story they found interesting.

Prepare to Read

Preview/Set Purpose Have students preview the book by looking at the headings, illustrations, and bulleted points. Have them explain how these text features will be helpful to them. Point out that this is a nonfiction book, which is a book about real people, places, and events. Ask them to predict what they will learn from this book.

Sequence

Teach/Model Display a sequence chart similar to the one below. Tell students that good readers pay attention to sequence, the order of events in a story. Ask students to describe what they do first, next, then, and last when they get up in the morning. Point out that authors often use time-order words to explain the order of a story.

First → **Next** → **Then** → **Last**

Practice/Apply Have partners record any time-order words they find in the selection. Add these to a class chart.

Summarize

Teach/Model Tell students that good readers stop at different points in a selection to summarize what they have read. Have them read pages 2–3 of the book. Say, *When I finish reading these pages, I ask myself: What was the overall point in that section? I also look at the heading and illustration to help me summarize. Using these features, I can summarize by saying that Roberto always dreamt of becoming a baseball player.*

Practice/Apply Have partners write a one-sentence summary after each section of the text.

Build Vocabulary

Help students use context to understand key vocabulary. Have them answer each question by reading the words and sentences that surround the word.

ballhawk (p. 6)	Who is a great *ballhawk*? **Possible response: Derek Jeter**
vanish (p. 6)	What feelings would you want to *vanish*? **anger, sadness**
fault (p. 10)	How do you feel when mistakes are your *fault*? **guilty, upset**
depend (p. 11)	What people do you *depend* on most in your life? **friends, family, teachers**
concentrate (p. 9)	Which school subject makes you *concentrate* most? **math**
outfielder (p. 6)	What are the qualities of a good *outfielder*? **focused, strong arm**

CM 79 **Independent Practice:** Vocabulary Practice page 79

Read and Comprehend

Monitor Comprehension

Guide students as they read the text, using the following questions:

Pages

2–5 What does Roberto do to help his dream come true? **He squeezes a ball and practices catching it.** IMPORTANT DETAILS

6–7 Why would American scouts come to watch Roberto? **They may have heard stories of his talent.** MAKE INFERENCES

8–11 What probably kept Roberto from quitting baseball? **He'd never been a quitter before.** DRAW CONCLUSIONS

What events led to winning the World Series? **Roberto learned how to swing at only good pitches.** SEQUENCE (Focus Skill)

12–14 Why do you think Roberto helped the needy? **He remembered being in that position.** SPECULATE

On page 14, what kind of person is Roberto? **generous** CHARACTERIZATION

15–16 What are the highlights of Roberto's career? **winning a World Series, MVP, Golden Gloves** SUMMARIZE (Focus Strategy)

What lessons could we learn from Roberto's life? **Never give up.** THEME

Think and Respond

Have students answer the questions on the inside back cover of the book.

1. **The first chapter tells of Roberto's younger years growing up. The second chapter tells of Roberto joining the Pirates. The third chapter tells of his last years in baseball.**
2. **Roberto joined the Pirates in 1955.**
3. **Possible response: It is called "Roberto's Dream" because Roberto's dream was to play baseball. The story is about how this dream came true.**
4. **Roberto Clemente is a hero because he was a great baseball player and he helped others in need.**
5. **Roberto's good fortune is that he was picked to play on the Pirates. His bad fortune is his back injury and his plane crash.**
6. **Possible response: I would tell him what a great man he was for helping the earthquake victims.**

Language Arts Connection Have students write an acrostic poem using all the letters in *Roberto.* This poem will have seven lines, each of which will relate to Roberto Clemente's life.

School-Home Connection Have students make a list of the qualities of a hero. Have them give examples of people they consider their heroes to support their point.

CM 80 **Independent Practice:** Comprehension Practice page 80

Write

Writer's Craft: Development

Write a Persuasive Letter Explain to students that development is about using strong reasons and examples to support opinions and ideas in writing. Explain that it is especially important when writing to persuade or convince. Turn to page 16 in the story. Read the highlights of Roberto Clemente's career. Have students select what they think was Roberto's most important accomplishment and list their reasons in a graphic organizer like the one below. Have them write a letter to persuade a friend of their opinion. Make sure they state their opinion and use three strong reasons and supporting examples to convince the reader accordingly.

Roberto's Greatest Accomplishment	→	Reason 1	→	Reason 2	→	Reason 3

Fluency

Oral Reading Fluency

CM 81

Assessment Use Oral Reading Fluency Practice page 81 to assess students' oral reading fluency.

Name ____________________ Roberto's Dream Vocabulary

▶ **Write the letter of the correct meaning in the blank.**

1. D outfielder
2. G ballhawk
3. B vanish
4. A concentrate
5. E fault
6. H depend

A to pay close attention
B to go suddenly out of sight; disappear
C a baseball player who is prone to injury
D a baseball player whose position is in the outfield
E responsibility for a mistake; blame
F to mistrust
G a baseball player who always seems to get to the ball and catch it
H to rely

▶ **Draw an award that you would give to Roberto Clemente. The award should contain reasons for giving the award. Use at least two vocabulary words in your explanation.**

Possible response: I am giving Roberto Clemente this award because he was a great ballhawk. Also, he was someone kind who you could really depend on.

© Harcourt

Books for All Learners Lessons, Grade 3 79

▲ Vocabulary Practice p. 79

Name ____________________ Roberto's Dream Comprehension

▶ **Respond to each of the following questions. Use complete sentences in your responses.**

1. Write two sentences to summarize Roberto Clemente's childhood.
   Possible response: Roberto was born in Puerto Rico. He was poor and had to work from a young age.
2. Write two sentences to summarize Roberto Clemente's baseball career.
   Possible response: Roberto went from the company baseball team, to the minor leagues to the major leagues. He was one of the only baseball players ever to make 3,000 hits.
3. Write two sentences to summarize how Roberto Clemente helped the needy.
   Possible response: Roberto helped the needy by donating food, clothing, and other things to them. He was going to help people hurt by an earthquake when he died in a plane crash.

▶ **Use the grid below to write a timeline of Roberto Clemente's baseball career.**

First	Next	Then	Finally
First Roberto joined his company baseball team.	Next, he joined the Minor Leagues in Puerto Rico.	Then, the Pittsburgh Pirates signed him.	Finally he became one of the greatest baseball players ever.

© Harcourt

80 Books for All Learners Lessons, Grade 3

▲ Comprehension Practice p. 80

Name ______________________________

▶ **Write the letter of the correct meaning in the blank.**

1. _____ outfielder
2. _____ ballhawk
3. _____ vanish
4. _____ concentrate
5. _____ fault
6. _____ depend

A to pay close attention

B to go suddenly out of sight; disappear

C a baseball player who is prone to injury

D a baseball player whose position is in the outfield

E responsibility for a mistake; blame

F to mistrust

G a baseball player who always seems to get to the ball and catch it

H to rely

▶ **Draw an award that you would give to Roberto Clemente. The award should contain reasons for giving the award. Use at least two vocabulary words in your explanation.**

____

____

____

Name ______________________________

▶ **Respond to each of the following questions. Use complete sentences in your responses.**

1. Write two sentences to summarize Roberto Clemente's childhood.

____

____

2. Write two sentences to summarize Roberto Clemente's baseball career.

____

____

3. Write two sentences to summarize how Roberto Clemente helped the needy.

____

____

▶ **Use the grid below to write a timeline of Roberto Clemente's baseball career.**

First	**Next**	**Then**	**Finally**

Name ______________________________

As a boy, Roberto carried a rubber ball around with him. He 12
squeezed it to make his hands stronger. At night, he threw a 24
baseball at the ceiling, catching it over and over. Of course, he 36
played baseball as much as he could. 43

Life wasn't easy for the Clemente family. Roberto's father 52
worked in the sugarcane fields. He helped harvest the sugarcane 62
that is made into sugar. He earned just $2 a week. 73

Later, Roberto's father became a foreman in charge of other 83
workers. As a foreman, he earned $4 a week. Still, he could not buy 97
his children many of the things they wanted. The money he earned 109
had to pay for things they needed. 116

When Roberto was nine, he wanted a bicycle. He asked his 127
father to buy one for him. His father said, "You'll have to earn it, 141
Roberto." 142

So, Roberto did. Each morning, he carried a neighbor's big, 152
empty milk pail half a mile to a store. After it was filled, 165
he carried it back. 169

FLUENCY SCORE

Total Words Read Per Minute ____________

Number of Errors − ____________

Number of Words Read Correctly ____________

The Subway Series

by David McCoy

SUMMARY Kevin is a baseball fan, his younger brother Andy is not. When their mother unexpectedly gets four tickets to Game 2 of the World Series at Yankee Stadium, Kevin is thrilled and insists that he invite a friend instead of having Andy go to the game. The boy's parents insist on going as a family, and Kevin learns some surprising things about his younger brother's talents.

- **Focus Skill:** Prefixes and Suffixes
- **Focus Strategy:** Use Decoding/Phonics
- **Vocabulary:** unexpected, glanced, prescription, comfort, contagious, longed, attention

Differentiated Instruction

Independent Reading Have students read the story independently. After each chapter, have them stop and write a short summary. Have students write down questions to ask if they do not understand a part of the story.

Prepare to Read

Preview/Set Purpose Have students read the title of the book and study the cover. Have students describe what they think the relationship between the boys must be and how the subway will play a part in the story.

Prefixes and Suffixes

Focus Skill

Teach/Model Write the following words on the board: *preview*, *preface*, *pretest*. Ask students to explain what the words have in common. Explain that each contains a prefix, a word part added to the beginning of a word that changes its meaning. Explain that a suffix is added to the end of a word and also changes the meaning of the word.

Practice/Apply Have students use the chart below to make five different terms. Explain that they can use any other prefixes and suffixes to create these words. Have students define each of their new words.

Prefix	Root	Suffix	Word/Definition
un-	move	-able	unmovable – not able to be moved
re-	new	-able	renewable – able to be made new again
un-	break	-able	unbreakable – not able to be broken

Use Decoding/Phonics

Focus Strategy

Teach/Model Tell students about the strategies that good readers use whenever they come to words they don't know. Explain that understanding the parts of a word, such as prefixes and suffixes, is one way to figure out a word's meaning. Display the word *decoding* and ask what it means. Have students find and define the prefix, root, and suffix of the word. Say, *When I come to a word I don't know, first I break it down by cutting off prefixes and suffixes and looking at the root or main part of the word to see if I recognize it. Then I add the meanings of the prefixes and suffixes to find a complete definition of the word. I could also ask myself if I know any other words with the same root.*

Practice/Apply Have students decode each of the following terms: *review*, *hopefully*, *disorganized*. Explain the methods students can use to decode each term and write the results on the board.

Build Vocabulary

Help students use context to understand key vocabulary. Have them answer each question by reading the words and sentences that surround the word.

unexpected (p. 3)	How would you decode *unexpected*? **After defining the prefixes and suffixes, I would think of other words with the same root.**
glanced (p. 3)	What word or phrase could you use in place of *glanced*? **looked, stared**
prescription (p. 16)	Where would you get a *prescription*? **doctor, pharmacist**
comfort (p. 10)	What or who can *comfort* you? **Possible response: my dog**
contagious (p. 11)	What different things could *contagious* describe? **laughter, disease**
longed (p. 4)	What birthday present have you *longed* for? **Possible response: a new computer**
attention (p. 6)	When did you really need *attention*? **Possible response: after a serious injury**

CM 85 **Independent Practice:** Vocabulary Practice page 85

Read and Comprehend

Monitor Comprehension

Guide students as they read the text, using the following questions:

Pages

2–5 What two events show the relationship between Kevin and Andy? **their arguments, Kevin wanting to exclude Andy from going to the game** CLASSIFY/CATEGORIZE

6–7 Which of the boys is right about the history of the song? **Andy** USE PRIOR KNOWLEDGE

8–10 Break the word *endangered* into word parts **en-danger-ed** PREFIXES AND SUFFIXES (Focus Skill)

11–13 What made Kevin start to change his mind about Andy? **Andy knows about baseball.** IMPORTANT DETAILS

14–16 How are the brothers alike and different? **Kevin is self-centered and hotheaded. Andy is more patient and clever. They come to understand each other.** COMPARE AND CONTRAST

What idea is the author trying to get across about siblings? **Siblings are not that bad if you give them a chance.** THEME

Think and Respond

Have students answer the questions on the inside back cover of the book.

1. **The first chapter tells about Kevin and Andy's parents getting World Series tickets; chapter two tells about the actual game; chapter three tells what happened on the ride home.**
2. **The prefix in the word impatiently is *im*, and the suffix is *ly*.**
3. **You should get to know and appreciate your sister or brother.**
4. **Kevin gives his brother the prized baseball.**
5. **Kevin begins to like his brother as the story progresses.**
6. **Possible response: I would have felt hurt because Kevin wanted to take a friend to the game instead of me. I would have felt better at the end because Kevin gave me the ball.**

Science Connection Have students make a list of facts they have heard about the bald eagle. Ask if they have seen one before.

School-Home Connection Think about the members of your family. Which one most reminds you of Andy? Describe this family member.

CM 86 **Independent Practice:** Comprehension Practice page 86

Write

Writer's Craft: Development

Write a Persuasive Review Explain that development is using strong reasons and examples to support opinions and ideas. For example, a well-developed paragraph that reviews a movie should first state an opinion and then support that opinion with convincing examples. Say, *If I want to convince you that the movie I saw was funny, I may explain that it made me laugh, and talk about some especially funny scenes. These examples will support my statement that the movie was funny.* Have students write a paragraph about the book *The Subway Series* in which they persuade someone to read the book. Have students use the graphic organizer below.

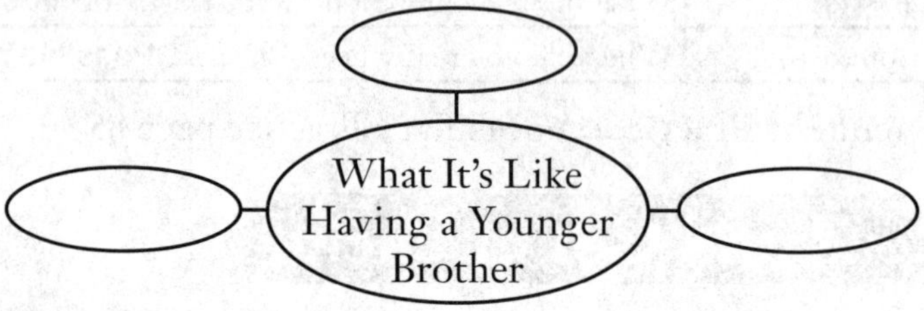

Fluency

Oral Reading Fluency

CM 87

Assessment Use Oral Reading Fluency Practice page 87 to assess students' oral reading fluency.

Name ______________________ The Subway Series Vocabulary

▶ **Circle the letter of the correct answer.**

1. The speaker had to scream to get the ________ of the crowd.
   (A) attention B prescription C unexpected D comfort
2. I ________ at the boy who kept throwing paper at me.
   A attention B unexpected C longed (D) glanced
3. Marlowe and I had always been close friends so his lie was ________.
   A longed B attention C contagious (D) unexpected
4. Karen was lonely as an only child and ________ for a sibling.
   A glanced (B) longed C attention D comfort
5. My brother was so ________ when he had the flu that he could not leave his room.
   (A) contagious B unexpected C prescription D comfort
6. When I am sick, all I want is quiet and ________ so that I can rest.
   (A) comfort B longed C unexpected D contagious
7. A headache was one side-effect of the ________ my doctor gave me.
   A contagious B attention C glanced (D) prescription

▶ **Write about a time you *longed* for something. Use two vocabulary words in your explanation.**

Every time I passed the bike shop, I glanced at the red scooter that got my attention last month.

© Harcourt

Books for All Learners Lessons, Grade 3 85

▲ Vocabulary Practice p. 85

Name ______________________ The Subway Series Comprehension

▶ **Find four conflicts the brothers face in the story. Then explain how each is resolved.**

Conflict or Problem	Resolution
Kevin wants to take a friend to the game.	Mom and Dad say no.
The game starts in just one hour.	They take the subway.
Kevin isn't sure when "The Star Spangled banner" was written.	Andy says it was written in 1814.
Kevin reaches past Andy to get the foul ball.	Kevin gives Andy the ball.

▶ **Complete the exercise below.**

Write a diary entry from Andy's point of view. Explain how he might have felt before and after the baseball game. Stay true to the events of the story.

Possible response: Dear Diary, Today started off as a disaster. Mom got tickets to the World Series, and I was so excited until Kevin tried to make me stay home and bring his friend instead. All day long he was mad at me. I couldn't say anything without him getting upset. He even snatched a ball from me that I was about to catch. But, by the end of the day, I guess it was okay. Kevin gave me back my ball.

© Harcourt

86 Books for All Learners Lessons, Grade 3

▲ Comprehension Practice p. 86

Name ______________________________

▶ **Circle the letter of the correct answer.**

1. The speaker had to scream to get the __________ of the crowd.

   A attention **B** prescription **C** unexpected **D** comfort

2. I __________ at the boy who kept throwing paper at me.

   A attention **B** unexpected **C** longed **D** glanced

3. Marlowe and I had always been close friends so his lie was __________.

   A longed **B** attention **C** contagious **D** unexpected

4. Karen was lonely as an only child and __________ for a sibling.

   A glanced **B** longed **C** attention **D** comfort

5. My brother was so __________ when he had the flu that he could not leave his room.

   A contagious **B** unexpected **C** prescription **D** comfort

6. When I am sick, all I want is quiet and __________ so that I can rest.

   A comfort **B** longed **C** unexpected **D** contagious

7. A headache was one side-effect of the __________ my doctor gave me.

   A contagious **B** attention **C** glanced **D** prescription

▶ **Write about a time you *longed* for something. Use two vocabulary words in your explanation.**

____

____

____

Name ____________________

▶ **Find four conflicts the brothers face in the story. Then explain how each is resolved.**

Conflict or Problem	Resolution

▶ **Complete the exercise below.**

Write a diary entry from Andy's point of view. Explain how he might have felt before and after the baseball game. Stay true to the events of the story.

____

____

____

____

____

____

____

Name ______________________________

The second game of the Subway Series! Kevin thought to himself. 11
And I'm here to see it. 17

The Yankees scored three runs in the first two innings. Neither 28
team scored during the next few innings. Then, in the bottom of the fifth 42
inning, Kevin's favorite Yankee, Paul O'Neill, hit a single, allowing the 53
Yankees to score again. Kevin jumped up and down. Even Andy was 65
jumping. The excitement was contagious. 70

After every play, Andy spouted some new fact. Kevin couldn't 80
believe how much Andy knew about baseball. Maybe his little brother 91
wasn't so bad after all. 96

It was the bottom of the seventh inning. Kevin's favorite player 107
was up at bat again. Ball one. Then he hit a foul ball high up into the 124
stands. The ball was headed straight for Andy, who was ready to catch 137
it. 138

At the last second, Kevin reached up and snatched the ball with 150
his mitt. A perfect catch. People cheered, and several adults sitting 161
behind Kevin congratulated him. 165

FLUENCY SCORE

Total Words Read Per Minute ____________

Number of Errors − ____________

Number of Words Read Correctly ____________

Poor Richard

by Stef Schumacher

SUMMARY Aside from being a statesman and inventor, Ben Franklin is known for publishing *Poor Richard's Almanack*, a collection of general information and wise sayings.

- **Focus Skill:** Narrative Elements
- **Focus Strategy:** Make and Confirm Predictions
- **Vocabulary:** summons, fortunate, generation, persistently, faithful, illuminated

Differentiated Instruction

Paired Reading Organize students in pairs. Have them read the selection aloud to each other, alternating with each page. After students have finished reading, have them discuss in pairs a part of the story they found interesting.

Prepare to Read

Preview/Set Purpose Have students examine the cover and title. Have students preview the book by looking at the illustrations and sections. Point out that this selection is a biography, or story of someone's life. Ask students what they already know about Ben Franklin and what they expect to learn from this book.

Focus Skill

Narrative Elements

Teach/Model Tell students that stories are made up of certain parts: setting, when or where the story takes place; characters, the people a story is about; a conflict, a problem that must be resolved; and a resolution, how the problem is solved. Display a chart similar to the one below. Tell students that good readers identify the parts of a story as they read. Work with students to have them identify the story that comprises this chart.

Setting	Characters	Conflict	Resolution
a poor farm	Jack – a young man Giant – an angry man	Jack's mother sends him to trade their cow for food. Jack trades the cow for magic beans instead.	Jack steals a goose that lays golden eggs to save his family from starvation.

Practice/Apply Have partners complete a chart like the one above after reading *Poor Richard.*

Focus Strategy

Make and Confirm Predictions

Teach/Model Tell students that good readers use clues like the book cover, title, and other text features to make a prediction, or thoughtful guess, about what will happen next in a story. Have them look at the cover. Say, *When I look at the cover I notice that the people are dressed very differently. Some people are in bonnets and interesting hats, and* shoppe *is spelled differently from our* shop. *I wonder if it will be like* Little House on the Prairie *or something similar.* Have students write their own predictions using the cover, title, and sections.

Practice/Apply Have partners confirm their predictions after reading by responding to the following questions: Were your predictions correct? How close was your guess? When did you realize that you were correct or not quite on target?

Build Vocabulary

Help students use context to understand key vocabulary. Have them answer each question by reading the words and sentences that surround the word.

summons (p. 7)	If a policeman *summons* you, how would you feel? **nervous, scared**
fortunate (p. 11)	Predict what might happen to a *fortunate* person. **winning the lottery**
generation (p. 16)	Who is a famous member of your grandparents' *generation*? **Mick Jagger**
persistently (p. 12)	Describe someone who *persistently* tries to get a job. **They do not give up.**
faithful (p. 14)	How would adding the prefix *un-* change the meaning of *faithful*? **It would mean the opposite.**
illuminated (p. 6)	How could your room be *illuminated*? **lights or lamps**

CM 91 **Independent Practice:** Vocabulary Practice page 91

Read and Comprehend

Monitor Comprehension

Guide students as they read the text, using the following questions:

Pages

2–3 What are some of Franklin's inventions that you have heard of or used yourself? **bifocals, lightning rod** CONNECT TO PERSONAL EXPERIENCE

4–7 If he did not attend school, how did Franklin learn? **He read and wrote a lot, asked questions, and learned from the world.** NARRATIVE ELEMENTS (Focus Skill)

Look at the picture on page 4. How are Franklin's books illuminated? **candle** USE GRAPHIC AIDS

8–11 If going to bed and getting up early make you wise, how would you describe a person who stays up late and oversleeps? **foolish, wastes time** COMPARE AND CONTRAST

12–15 Who has set an example for you? **parents, teachers** EXPRESS PERSONAL OPINION

16 What were Franklin's most important contributions? **inventions, writing** ANALYZE

How could you convince people that *haste makes waste*? **Instead of making errors you will have to fix, take your time and do it right the first time.** SPECULATE

What do you think Franklin was most proud of accomplishing in his life? **the popularity of *Poor Richard's Almanack*** MAKE PREDICTIONS (Focus Strategy)

Think and Respond

Have students answer the questions on the inside back cover of the book.

1. **The writer explains each saying and gives examples.**
2. **Possible response: The writer begins the book with information about Ben Franklin so the reader will know who he was.**
3. **They have remained popular because they are still true today.**
4. **Possible response: Franklin wrote *Poor Richard's Almanack* to help people.**
5. **Possible response: They were alike because they both produced their own newspapers and invented useful things. They were different because Franklin helped with the government and Edison did not.**
6. **Possible response: I like the saying, "No gains without pains." This advice is helpful because it teaches me to work hard to do something well.**

Language Arts Connection Have students make a list of items that they think should be included in an almanac.

School-Home Connection Have students select a close friend or relative who makes them think of one of Franklin's famous sayings. Have them explain their choice.

CM 92 **Independent Practice:** Comprehension Practice page 92

Write

Writer's Craft: Development

Write a Persuasive Letter Work with students to write a persuasive letter with good development from a writing prompt. Say, *Read the prompt to find out what information you will need. Ask yourself these questions: What is the writing form, who is the audience, what is the topic, and does the prompt have any special instructions? Then, organize your ideas using a graphic organizer such as an idea web.* Have students use the following as their writing prompt: *Write a response to the letter below. Imagine you are giving the advice Ben Franklin himself might have given.* Work with students in a prewriting exercise to develop reasons why school is important.

> Dear Friend,
>
> Although I work all the time, I have nothing to show for it. Every day it is harder to get up because I stay up late with my friends. I don't want to give up my friends and fun, but I don't know what else to do. Help!
>
> Sincerely,
>
> Lost in Labor

Fluency

Oral Reading Fluency

CM 93

Assessment Use Oral Reading Fluency Practice page 93 to assess students' oral reading fluency.

Name ____________________

Poor Richard
Vocabulary

▶ Write the letter of the word that best completes each analogy below.

1. denies: rejects :: summons: **b**
   a) brings **b)** accepts **c)** eats **d)** grows
2. angrily: worked :: persistently: **d**
   a) draw **b)** saw **c)** scream **d)** talked
3. lamp: illuminated :: fridge: **c**
   a) cooked **b)** ate **c)** cooled **d)** froze
4. hot: cold :: faithful : **d**
   a) loyal **b)** true **c)** unsure **d)** disloyal
5. band: musician :: generation: **a**
   a) grandparent **b)** age **c)** year **d)** time

▶ Write a thank-you letter to Ben Franklin to honor his most important accomplishments. Use at least two vocabulary words.

Possible response: Dear Mr. Franklin, Thank you for being the greatest man of your generation. You illuminated many things for us.

▲ Vocabulary Practice p. 91

Name ____________________

Poor Richard
Comprehension

▶ Explain each quotation in your own words.

Quotations from *Poor Richard*	My thoughts
Lost time is never found again. (p. 10)	**Don't waste time.**
When the well's dry, we know the worth of water. (p. 11)	**Appreciate what you have.**
No gains without pains. (p. 12)	**Good work is hard work.**
Speak little; do much. (p. 13)	**Don't talk about things, do them.**
A true friend is the best possession. (p. 14)	**Nothing's better than friends.**

▶ Explain how the author showed the character of Ben Franklin.

Possible response: The author showed how Ben Franklin had great accomplishments like inventing bifocals and writing famous sayings. He did all that and didn't even get to go to school like we do. Also, he was always working so he didn't waste time.

▲ Comprehension Practice p. 92

Name ______________________________

▶ **Write the letter of the word that best completes each analogy below.**

1. denies: rejects :: summons: _____
a) brings **b)** accepts **c)** eats **d)** grows

2. angrily: worked :: persistently: _____
a) draw **b)** saw **c)** scream **d)** talked

3. lamp: illuminated :: fridge: _____
a) cooked **b)** ate **c)** cooled **d)** froze

4. hot: cold :: faithful : _____
a) loyal **b)** true **c)** unsure **d)** disloyal

5. band: musician :: generation: _____
a) grandparent **b)** age **c)** year **d)** time

▶ **Write a thank-you letter to Ben Franklin to honor his most important accomplishments. Use at least two vocabulary words.**

____

____

____

Name ____

▶ **Explain each quotation in your own words.**

Quotations from *Poor Richard*	My thoughts
Lost time is never found again. (p. 10)	
When the well's dry, we know the worth of water. (p. 11)	
No gains without pains. (p. 12)	
Speak little; do much. (p. 13)	
A true friend is the best possession. (p. 14)	

▶ **Explain how the author showed the character of Ben Franklin.**

____

____

____

____

____

Build Vocabulary

Help students use context to understand key vocabulary. Have them answer each question by reading the words and sentences that surround the word.

colonel (p. 15)	Would a *colonel* be highly ranked in an army? **Yes.**
soldiers (p. 15)	What qualities do *soldiers* have? **bravery, leadership**
urgent (p. 5)	Give an example of an *urgent* problem. **Possible response: a house on fire**
brambles (p. 8)	What kinds of plants have *brambles*? **rose bushes, holly trees**
stumbling (p. 15)	What would cause a person's *stumbling*? **Possible response: being tired**
outstretched (p. 15)	Why would you keep a hand *outstretched*? **Possible response: to help someone up**
weary (p. 15)	What activity makes you *weary*? **Possible response: running, doing chores**

CM 97 **Independent Practice:** Vocabulary Practice page 97

Read and Comprehend

Monitor Comprehension

Guide students as they read the text, using the following questions:

Pages

2–4 Why did Josh's dad start telling him the story? **because Josh was upset about not having anyone famous in his family** MAKE INFERENCES

5–8 How was Maggie's home in Nebraska different from her home in St. Louis? **It was made of sod instead of wood.** COMPARE AND CONTRAST

9–11 What do you think is going to happen next? **Maggie's father will get stuck in the snow.** MAKE PREDICTIONS

12–13 Summarize pages 12–13. **There is a snowstorm and the family is worried about Papa, but they try not to worry by having school and singing.** SUMMARIZE (Focus Skill)

14–16 If you didn't understand how Papa returned, which page would you reread? **page 15** REREAD TO CLARIFY (Focus Strategy)

What is the main idea of the story? **Possible response: Everyone's family has interesting people in it, whether they are famous or not.** MAIN IDEA AND DETAILS

Think and Respond

Have students answer the questions on the inside back cover of the book.

1. **Maggie, Lizzie, Baby Nate, Mama, and Papa Davis are from the past. Josh and his dad are from the present. The author says that Josh and his dad are reading the journal, which was written in 1850.**
2. **Possible response: Josh is upset that there are no famous people in his family. Josh and his dad read the journal and Josh feels better about his history.**
3. **They lived in a sod house far from stores and schools, and they worked very hard.**
4. **Possible response: Pioneer life was hard because there were no cars or electricity.**
5. **Possible response: My house is the same because it has walls. It is different because hers was made of dirt and grass, and mine is made of wood and brick.**
6. **Possible response: I would not have liked being a pioneer in 1850 because they didn't have electricity.**

SOCIAL STUDIES

Social Studies Connection Ask students what they know about pioneer life and what books, web sites, or films they might examine to find out more.

School-Home Connection Have students sit down with a member of their family and create a family tree to share with the class.

CM 98 **Independent Practice:** Comprehension Practice page 98

Write

Writer's Craft: Effective Sentences

Write a Paragraph that Compares Say, *An example of an effective topic sentence for a paragraph that compares Martin Luther King, Jr. and George Washington would be: Even though Martin Luther King and George Washington seem very different, they actually have a lot in common.* Explain to students that good writers use a variety of sentence structures and lengths. One type of paragraph compares two things, or tells how they are alike. Have students use the chart below to write a paragraph that compares two heroes. Remind them to use a topic sentence, supporting details, and a closing sentence in their paragraph.

Hero 1	Both Heroes	Hero 2
Possible response: Batman uses fancy gadgets and a cool car.	**Possible response: Batman and Superman save people.**	**Possible response: Superman uses his strength and flies.**

Fluency

Oral Reading Fluency

CM 99

Assessment Use Oral Reading Fluency Practice page 99 to assess students' oral reading fluency.

Name ______________________

The Best Thanksgiving
Vocabulary

▶ **Fill in the blank with the correct word.**

outstretched	urgent	colonel	weary
stumbling	brambles	soldiers	

1. The **colonel** gave orders to the army.
2. I was **weary** after playing basketball all day.
3. The **soldiers** have a lot of courage.
4. I hurt my finger on the **brambles** of the rosebush.
5. The fire department went on an **urgent** call.
6. The man started **stumbling** down the hall.
7. My mother hugged me with **outstretched** arms.

▶ **Draw a picture of an *urgent* problem. Then, write a caption using at least four vocabulary words.**

Caption: **Possible response: The stumbling soldiers looked to the weary colonel for help.**

Books for All Learners Lessons, Grade 3 97

▲ Vocabulary Practice p. 97

Name ______________________

The Best Thanksgiving
Comprehension

▶ **Fill in the gaps to complete the summary of *The Best Thanksgiving.***

Josh was upset because there was nobody **famous** (1) in his family. His dad told him a story about his **great-great-grandmother** (2) Maggie Davis. She kept a **journal** (3) about her life on the Nebraska **prairie** (4). Maggie wrote about her **father** (5) going to Ft. Kearney for **supplies** (6). He was gone for a **long** (7) time. Maggie and her mother and siblings had fun by **Possible response: telling stories** (8) and **Possible response: polishing silver** (9).

▶ **Explain how rereading for clarity helped you understand the time shifts in *The Best Thanksgiving.***

***The Best Thanksgiving* was hard to understand because it kept switching back and forth between past and present. There was a diary and a story happening at the same time. I went back and read sections that I didn't understand. It helped me understand the story better. I understood when things were going on in the past or present.**

98 Books for All Learners Lessons, Grade 3

▲ Comprehension Practice p. 98

Name ______________________________

Fill in the blank with the correct word.

outstretched	**urgent**	**colonel**	**weary**
stumbling	**brambles**	**soldiers**	

1. The ______________________ gave orders to the army.
2. I was ______________________ after playing basketball all day.
3. The ______________________ have a lot of courage.
4. I hurt my finger on the ______________________ of the rosebush.
5. The fire department went on an ______________________ call.
6. The man started ______________________ down the hall.
7. My mother hugged me with ______________________ arms.

Draw a picture of an *urgent* problem. Then, write a caption using at least four vocabulary words.

Caption: ______________________________

____

____

Name ______________________________

▶ Fill in the gaps to complete the summary of *The Best Thanksgiving.*

Josh was upset because there was nobody ________ (1) in his family. His dad told him a story about his ________ (2) Maggie Davis. She kept a ________ (3) about her life on the Nebraska ________ (4). Maggie wrote about her ________ (5) going to Ft. Kearney for ________ (6). He was gone for a ________ (7) time. Maggie and her mother and siblings had fun by ________ (8) and ________ (9).

▶ Explain how rereading for clarity helped you understand the time shifts in *The Best Thanksgiving.*

____

____

____

____

____

____

____

Name ______________________________

Afternoon (I think), November 25, 1850 **6**

Mama was right. It has been snowing for two days. It is snowing so hard **21**
you can't see anything but white. Mama went out to tend the animals. The **35**
wind was blowing so hard that when she came back, we could hardly shut **49**
the door. Mama sang songs with us. It almost made us forget Papa. **62**

Early morning, November 26, 1850 **67**

There is still no sign of Papa, only snow and more snow. The wind **81**
howled around the house all night. We couldn't sleep. Lizzie thought the **93**
howling was a pack of wolves. She was crying and howling herself until **106**
Mama said, "That's enough, Lizzie. Let's have a play party!" Mama played **118**
as though she was a fiddler and sang songs. Lizzie and I danced around with **133**
each other and with the baby. **139**

Afternoon, November 27, 1850 **143**

The wind has died down and the snow has stopped. Papa is still not here. **158**
We had school, and then Mama told us fairy tales. **168**

FLUENCY SCORE

Total Words Read Per Minute ____________

Number of Errors − ____________

Number of Words Read Correctly ____________

Realistic Fiction

Friends Forever

by Rebecca Motil

SUMMARY Megan and Sue Ann meet at camp. They find that they have many things in common. At camp they swim, hike, and make crafts together. They also spend every night learning to use a star map to locate stars. By the end of camp they have learned all about stars and about each other.

- **Focus Skill:** Compare and Contrast
- **Focus Strategy:** Summarize
- **Vocabulary:** canyon, pride, swiftly, skillful, feast, gazing, arranged

Differentiated Instruction

Paired Reading Organize students in pairs. Have them read the selection aloud to each other, alternating with each page. After students have finished reading, have them discuss in pairs a part of the story they found interesting.

Prepare to Read

Preview/Set Purpose Have students examine the cover and title. Have students preview the book by looking at the illustrations and star maps. Ask them what they think the book might be about. Point out that this is a realistic fiction book about made-up characters who do real-life things. Ask them what they think they will learn from this book.

Focus Skill

Compare and Contrast

Teach/Model Explain to students that when you compare things, you tell how they are alike. Then explain that when you contrast things, you tell how they are different. Display a chart similar to the one below. Say, *Good readers pay attention to the way characters are alike and different in order to understand a story.* Ask students to give examples of ways that they are alike and different from their best friend. Have students read pages 8–10 of *Friends Forever* and fill out the chart with one way Megan and Sue Ann are like Claire and one way they are different.

Megan and Sue Ann	All	Claire
Megan and Sue Ann don't know much about stars.	The three girls all like sleep-away camp.	Claire knows a lot about stars.

Practice/Apply Have partners add to their charts after they read *Friends Forever.*

Focus Strategy

Summarize

Teach/Model Explain to students that when you summarize what you have read, you state what has happened in your own words. Tell students that stories might have many characters who do different things. It helps to stop reading every few pages and summarize what has happened so far. Have students read pages 2–4 of *Friends Forever.* Say, *I could summarize these pages by saying that Megan was excited about going to sleep-away camp, but she was nervous when her parents left. Then she met her group leader Claire and her bunkmate Sue Ann.*

Practice/Apply Have partners stop after every three pages and summarize the events of that section of the book.

Build Vocabulary

Help students use context to understand key vocabulary. Have them answer each question by reading the words and sentences that surround the word.

canyon (p. 13)	Name a famous *canyon.* **the Grand Canyon**
pride (p. 13)	What do you call people who take *pride* in themselves? **proud**
swiftly (p. 8)	What word or phrase can you use in place of *swiftly*? **quickly, fast**
skillful (p. 13)	In what sport is Tiger Woods most *skillful*? **golf**
feast (p. 6)	How do you feel after you have a *feast*? **Possible response: full, sleepy**
gazing (p. 8)	How would you feel if someone kept *gazing* at you? **uncomfortable**
arranged (p. 5)	What word can you use in place of *arranged*? **organized**

CM 113 **Independent Practice:** Vocabulary Practice page 113

Read and Comprehend

Monitor Comprehension

Guide students as they read the text, using the following questions:

Pages

2–4 Why did Megan's feelings about her parents leaving change? **She became nervous about them not being there.** DRAW CONCLUSIONS

5–7 Why does Megan say "too bad you don't like corn" to Sue Ann? **She is making a joke because Sue Ann ate three ears of corn.** INTERPRETING INFORMATION

8–9 What is the other name for the North Star shown in the picture on page 9? **Polaris** LOCATE INFORMATION

10–13 Look at the star map on page 11. Where do you think the constellation Draco the Dragon got its name? **It looks like a dragon with a head and a tail.** USE GRAPHIC AIDS

14–16 What lesson did Megan learn at the end of the story? **She learned that her friendship with Sue Ann would last forever just like the stars.** SUMMARIZE (Focus Strategy)

How will Megan's nights at home be similar to her nights at camp? How will they be different? **She will look at the stars in both places, but she won't be with Sue Ann at home.** COMPARE AND CONTRAST (Focus Skill)

Think and Respond

Have students answer the questions on the inside back cover of the book.

1. **The story takes place at sleep-away camp.**
2. **No. At the beginning of the story, Megan saw the stars as pretty diamonds. At the end of the story, the stars reminded her of camp and her new friend Sue Ann.**
3. **It teaches that true friendship doesn't change.**
4. **Possible response: I learned some of the summer constellations.**
5. **Megan and Sue Ann were alike because they liked and disliked the same foods. They were different because Megan was a good swimmer and Sue Ann was not.**
6. **Possible response: I would like to ride my bike with my new friend.**

Science Connection Have the students look up the North Star on a star map. Have them research how the North Star is used for finding directions.

School-Home Connection List observatories in your area. Have students write a letter asking a family member to take them to one.

CM 114 **Independent Practice:** Comprehension Practice page 114

Write

Writer's Craft: Effective Sentences

Write a Paragraph that Contrasts Say, *Here is an example of an effective topic sentence for a paragraph that contrasts swimming and hiking: You can see that the two activities, swimming and hiking, are very different.* Explain to students that good writers use a variety of sentence structures and lengths. One important type of paragraph contrasts two things. Have students write a paragraph contrasting two summer activities. Remind them to use a topic sentence, supporting details, and a closing sentence in their paragraph.

Fluency

Oral Reading Fluency

CM 115

Assessment Use Oral Reading Fluency Practice page 115 to assess students' oral reading fluency.

Name ______________________

Friends Forever
Vocabulary

▶ **For each pair of words, put an "O" in the blank if the words have opposite meanings, and an "S" in the blank if the words mean the same.**

O 1. *pride*, shame

O 2. *canyon*, mountain

S 3. *swiftly*, quickly

O 4. *skillful*, clumsy

S 5. *arranged*, organized

S 6. *gazing*, staring

S 7. *feast*, party

▶ **Write a paragraph about summer camp using four vocabulary words.**

Possible response: My parents arranged for me to attend a summer camp near Big Canyon Falls. I was not very skillful, and everyone else learned things more swiftly.

▲ Vocabulary Practice p. 103

Name ______________________

Friends Forever
Comprehension

▶ **Circle the best answer for each question below.**

1. Which can be seen in the sky over the United States on July 15?
   - **A** Orion the Hunter
   - **B** The Big Dipper
   - Ⓒ Polaris (the North Star)
   - **D** Cancer the Crab

2. Why was the North Star important to Christopher Columbus?
   - Ⓕ He used it to cross the Atlantic Ocean.
   - **G** He thought it was pretty.
   - **H** It was the brightest star in the sky.
   - **J** It helped him tell time.

3. Which of these statements about Megan and Sue Ann is *false*?
   - **A** They both love pizza.
   - Ⓑ They are both good swimmers.
   - **C** They are both in the same cabin.
   - **D** They both like corn on the cob.

▶ **Explain how the pictures on pages 12 and 13 are alike and different.**

Possible response: Both show children having a good time, exercising, and taking part in camp activities. Page 12 shows children hiking and painting. Page 13 has children swimming and paddling a boat.

▲ Comprehension Practice p. 104

Name ______________________________

For each pair of words, put an "O" in the blank if the words have opposite meanings, and an "S" in the blank if the words mean the same.

___ **1.** *pride*, shame

___ **2.** *canyon*, mountain

___ **3.** *swiftly*, quickly

___ **4.** *skillful*, clumsy

___ **5.** *arranged*, organized

___ **6.** *gazing*, staring

___ **7.** *feast*, party

Write a paragraph about summer camp using four vocabulary words.

Name ____

▶ Circle the best answer for each question below.

1. Which can be seen in the sky over the United States on July 15?

A Orion the Hunter
B The Big Dipper
C Polaris (the North Star)
D Cancer the Crab

2. Why was the North Star important to Christopher Columbus?

F He used it to cross the Atlantic Ocean.
G He thought it was pretty.
H It was the brightest star in the sky.
J It helped him tell time.

3. Which of these statements about Megan and Sue Ann is *false*?

A They both love pizza.
B They are both good swimmers.
C They are both in the same cabin.
D They both like corn on the cob.

▶ Explain how the pictures on pages 12 and 13 are alike and different.

____

____

____

____

____

____

____

Name ______________________________

When they returned to the cabin, Claire showed all the girls her 12
star map. She pointed out the Summer Triangle and another 22
constellation called Draco the Dragon. 27

"Here's one you know," said Claire. She pointed to the North 38
Star. "Christopher Columbus used the North Star to help him cross 49
the Atlantic Ocean. It showed him where north was. It still shows 61
us north," Claire said. 65

The two weeks at camp passed swiftly. The campers went on 76
hikes. Sometimes they hiked in the woods. Once they hiked past a 88
canyon. 89

Every day they went swimming in the lake. Megan was a good 101
swimmer. She took pride in her swimming. Sue Ann was not as 113
skillful at swimming. Megan tried to help her as much as she could. 126

Each night, they looked up at the starry sky. They used Claire's 138
star map to find special stars and constellations. They found the 149
Big Dipper and the North Star. 155

On their last night at camp, it was cloudy. 164

FLUENCY SCORE

Total Words Read Per Minute ____________

Number of Errors − ____________

Number of Words Read Correctly ____________

Folktale

Why Butterflies Were Made

by Vita Jimenez

SUMMARY As a young girl and her grandfather watch butterflies in the park, Grandpa tells the tale of two Native American girls named Sun and Moon. Their story explains why butterflies were made.

- **Focus Skill:** Summarize
- **Focus Strategy:** Make and Confirm Predictions
- **Vocabulary:** satisfied, duty, council, mischief, nonsense, tidbits

Differentiated Instruction

Organize students into pairs. Have them read the selection aloud to each other, alternating with each page. After students have finished reading, have them discuss in pairs a part of the selection they found interesting.

Prepare to Read

Preview/Set Purpose Have students examine the cover and title. Have students preview the book by looking at the illustrations. Ask them where they think the story takes place. Point out that this is a folktale, which is a story that explains where things come from or why they are the way they are. Ask them what they think they will learn from this book.

Focus Skill

Summarize

Teach/Model Remind students that to summarize a story is to retell the important ideas in their own words. Tell students that good readers summarize events to make sure they understand a story. Have students read pages 2–5. Say, *I would summarize this section by saying that Sun and Moon were twin sisters who made everyone in their tribe happy.* Remind students that they should retell events in the order they happen in the story. Students may write keywords to help them summarize.

Practice/Apply Have partners summarize each section (every three pages) after they read *Why Butterflies Were Made.*

Focus Strategy

Make and Confirm Predictions

Teach/Model Remind students that good readers often make predictions, or guesses, about what happens next in the story. They use story clues and what they already know to guess what might happen next. Then they confirm, or check, their guesses by reading the story. Good readers think about what they read as they are reading. Say, *If I made a prediction based on pages 2–5, I would say that Sun and Moon are going to have an adventure. I say this because they are the main characters of the story and main characters usually have adventures.*

Practice/Apply Have partners stop at the end of each section (every three pages) of *Why Butterflies Were Made* and make predictions about what might happen next in the story.

Build Vocabulary

Help students use context to understand key vocabulary. Have them answer each question by reading the words and sentences that surround the word.

satisfied (p. 16)	What word could you use in place of *satisfied*? **Possible response: content, happy**
duty (p. 16)	What is the *duty* of a police officer? **to protect people**
council (p. 12)	Who might come to a *council* meeting? **Possible response: wise people, older people**
mischief (p. 4)	Who is a cartoon character who creates *mischief*? **Possible response: SpongeBob**
nonsense (p. 10)	What word means the opposite of *nonsense*? **Possible response: serious**
tidbits (p. 13)	What parts of an hour would be *tidbits*? **minutes, seconds**

CM 109

Independent Practice: Vocabulary Practice page 109

Read and Comprehend

Monitor Comprehension

Guide students as they read the text, using the following questions:

Pages

2–3 Why do you think Grandpa told the story? **Possible response: because his granddaughter loved looking at butterflies** MAKE INFERENCES

4–5 Summarize the reasons the tribe loved Sun and Moon? **They helped their parents, they kept the village happy, and they never got into mischief.** SUMMARIZE

6–7 What do Sun and Moon do with the baby bird? **They dig a hole near the cactus and bury it.** MAIN IDEA AND DETAILS

8–11 Do you think Sun will ever get out of bed? **Possible response: Yes.** MAKE PREDICTIONS

What does Sun's father think about her staying in bed? **He thinks she is being silly.** INTERPRETING INFORMATION

12–16 What do the elders decide to do after Father talks with them? **They decide to try to make Sun happy.** IMPORTANT DETAILS

Think and Respond

Have students answer the questions on the inside back cover of the book.

1. **Possible response: It is different because it introduces another story, the one Grandpa tells.**
2. **Butterflies were made to cheer up a sad little girl.**
3. **This story shows that the bond between the sisters and their parents is a close one.**
4. **Possible response: Some Native American folktales explain things in nature.**
5. **Sun and Moon are different from the butterflies because they are humans and the butterflies are insects. They are the same because butterflies have two wings that look alike just as Sun and Moon look alike.**
6. **Possible response: I would have created a puppy to make her smile again.**

Social Studies Connection Have students list questions about the Papago Indians.

School-Home Connection Have students read *Why Butterflies Were Made* to someone at home.

CM 110

Independent Practice: Comprehension Practice page 110

Write

Writer's Craft: Effective Sentences

Write a Paragraph that Explains Say, *In the story, Grandfather explains to his granddaughter how butterflies were made. You are going to write a paragraph to explain how your favorite sandwich is made.* Explain to students that good writers use a variety of sentence structures and lengths. Point out that good expository writing includes details that explain how and/or why things happen. Have students write a paragraph explaining the steps in making a sandwich. Remind them to use a topic sentence, supporting details, and a closing sentence.

Fluency

Oral Reading Fluency

CM 111

Assessment Use Oral Reading Fluency Practice page 111 to assess students' oral reading fluency.

Name ____________

Why Butterflies Were Made
Vocabulary

▶ Complete each analogy with the correct word.

duty council tidbits nonsense mischief satisfied

1. game : players :: meeting : **council**
2. hot : cold :: seriousness : **nonsense**
3. slim : skinny :: troublemaking : **mischief**
4. fat : plump :: content : **satisfied**
5. sad : gloomy :: job : **duty**

▶ Write a paragraph about a time when you or someone you know was caught making mischief. Use *tidbits* and two other vocabulary words.

Possible response: My friend Ty was not satisfied until he found mischief. We heard some tidbits about this old haunted house. Ty thought it was our duty to check it out. Sadly, our parents caught us before we crossed the street.

▲ Vocabulary Practice p. 109

Name ____________

Why Butterflies Were Made
Comprehension

▶ Circle T if the statement is true and F if it is false. If the statement is false, rewrite it to make it true.

1. (T) F Sun becomes sad because she and Moon find a dead baby bird.
2. T (F) Sun laughs and claps at the Elders' dancing.
Their dancing makes her sadder.
3. (T) F Wise Elder makes Sun happy again by making butterflies.
4. T (F) Sun's father asks the council to bring the bird back to life.
He asked them to make Sun happy.
5. (T) F Sun's mother is very worried when she won't get out of bed.

▶ Explain what you think Sun did the next time she felt sad.

Possible response: Sun thought about butterflies.

▲ Comprehension Practice p. 110

Name ____________________

Complete each analogy with the correct word.

duty	council	tidbits	nonsense	mischief	satisfied

1. game : players :: meeting : ____________
2. hot : cold :: seriousness : ____________
3. slim : skinny :: troublemaking : ____________
4. fat : plump :: content : ____________
5. sad : gloomy :: job : ____________

Write a paragraph about a time when you or someone you know was caught making mischief. Use *tidbits* and two other vocabulary words.

Name ____

▶ **Circle T if the statement is true and F if it is false. If the statement is false, rewrite it to make it true.**

1. T F Sun becomes sad because she and Moon find a dead baby bird.

____

2. T F Sun laughs and claps at the Elders' dancing.

____

3. T F Wise Elder makes Sun happy again by making butterflies.

____

4. T F Sun's father asks the council to bring the bird back to life.

____

5. T F Sun's mother is very worried when she won't get out of bed.

____

▶ **Explain what you think Sun did the next time she felt sad.**

____

____

____

____

Name ____

Long ago, there lived a very wise and kind man. People **11**
loved him because he was always fair and just. The man and his **24**
wife had twin girls. One was named Sun and the other was named **37**
Moon. The two girls were alike in every way. They had long, **49**
brown hair. Their eyes were the color of the night sky. Their smiles **62**
were like sunlight. **65**

The two girls dressed the same, too. If one wore flowers, the **77**
other did the same. The girls were kind to everyone, and they never **90**
made any mischief. **93**

People said that the girls kept the village happy. They were **104**
certainly a joy to their father. Sometimes, when he went away, they **116**
would give him tiny white stones. He would look at the stones and **129**
know that his daughters were thinking of him. They would run to **141**
meet him when he came back. **147**

They were a joy to their mother, too. They gave her beautiful **159**
bird feathers that they found and helped her in the garden. **170**

FLUENCY SCORE

Total Words Read Per Minute ____________

Number of Errors − ____________

Number of Words Read Correctly ____________

Wolf Tales

by Meish Goldish

SUMMARY: Three young wolf pups learn all about their famous family tree in bedtime stories read to them by their father.

- **Focus Skill:** Compare and Contrast
- **Focus Strategy:** Summarize
- **Vocabulary:** delighted, dusk, embraced, latch, tender, brittle, cunning

Differentiated Instruction

Paired Reading Organize students in pairs. Have them read the selection aloud to each other, alternating with each page. After students have finished reading, have them discuss in pairs a part of the story they found interesting.

Prepare to Read

Preview/Set Purpose Have students preview the table of contents and the illustrations. Remind students that fiction stories have characters and plots that are made up. Ask them what they think the book is about and why they might want to read it.

Focus Skill

Compare and Contrast

Teach/Model Remind students that good readers pay attention to similarities and differences in a text. Tell students that describing how things are alike and how they are different is called compare and contrast. Display a chart like the one below. Work with students to use the picture on page 2 to compare and contrast Mama and Papa Wolf. Tell students that a chart like this can also help them keep track of what they read. It will help them better understand what the stories have in common and how the stories are different.

Mama Wolf	Both	Papa Wolf
sitting alone in a rocking chair	reading	sitting on a bed with pups

Practice/Apply After reading, have partners list the similarities and differences between the first two stories.

Focus Strategy

Summarize

Teach/Model Remind students that to summarize is to retell the main events from a story in order. Tell them that a summary should be brief but still have all of the most important events from a story. Remind students that summarizing can help them better remember and understand what happens in a story. Have students look at pages 4–7. Say, *I would summarize* "Peter and the Wolf" *in 3 sentences: Great-Uncle Wolf lives in the forest. A boy named Peter sees Great-Uncle Wolf swallow a duck. Peter captures Great-Uncle Wolf, saves the duck, and takes Great-Uncle Wolf to the zoo.*

Practice/Apply After they have read the text, have partners write a summary of the events in this story.

Build Vocabulary

Help students use context to understand key vocabulary. Have them answer each question by reading the words and sentences that surround the word.

delighted (p. 2)	What word means the opposite of *delighted*? **unhappy, sad**
dusk (p. 6)	Which meal do you eat at *dusk*? **dinner**
embraced (p. 12)	If you have *embraced* someone, what have you used? **your arms**
latch (p. 14)	If you *latch* the door, what are you doing? **You are locking it.**
tender (p. 6)	Why would a fawn be described as *tender*? **Tender means soft and gentle.**
brittle (p. 14)	Why must you be careful with things that are *brittle*? **They break easily.**
cunning (p. 10)	What might a *cunning* person try to do? **He or she may try to fool or trick you.**

CM 115 **Independent Practice:** Vocabulary Practice page 115

Read and Comprehend

Monitor Comprehension

Guide students as they read the text, using the following questions:

Pages

2–3 Why don't the pups want to go to bed? **They are not tired and they want to hear a wolf story.** MAKE INFERENCES

4–7 What did Great-Uncle Wolf do that made Peter upset? **He swallowed a whole duck.** NOTE DETAILS

8–11 Which storyteller told the tale of "A Wolf in Sheep's Clothing"? **Aesop** NOTE DETAILS
How are wolves and sheep alike and different? **Possible response: Both have four legs and live outside; wolves eat meat while sheep eat grass.** COMPARE AND CONTRAST (Focus Skill)

12–14 Summarize Cousin Wolfy's plan. **Possible response: He will pose as the grandmother to get Little Red Riding Hood's cookies.** SUMMARIZE (Focus Strategy)

15–16 Do you think Cousin Wolfy's plan was a good one? **No, because he didn't get the cookies.** EXPRESS PERSONAL OPINION

Think and Respond

Have students answer the questions on the inside back cover of the book.

1. **It lists the title of each tale and the page number it begins on.**
2. **The three tales are alike because the main character is a wolf in each one. They are different because sometimes the human tricks the wolf and sometimes the wolf tricks the human.**
3. **This book is about new versions of famous tales about wolves, told from a wolf's viewpoint.**
4. **Possible response: I learned that the wolves are not as mean as people think.**
5. **The humans are alike because they beat the wolf in each story. They are different because one takes the wolf to the zoo, one captures the wolf by accident, and one gets away from the wolf.**
6. **Possible response: My favorite is "Little Red Riding Hood." I like it because it is funny.**

Language Arts Connection Have pairs act out one of the stories from *Wolf Tales*. Have them draw a program for their play.

School-Home Connection Have students ask family members or friends if they know a story about wolves, or any other wild animal.

CM 116 **Independent Practice:** Comprehension Practice page 116

Write

Writer's Craft: Effective Sentences

Write a Compare and Contrast Essay Have students read the third paragraph on page 6 of *Wolf Tales*. Say, *This paragraph is a good example of how the author uses a variety of sentences. The first sentence is short, while the second sentence is longer. This helps make the sentences more interesting.* Tell students that they will be writing an essay to compare and contrast the wolves in *Wolf Tales* with real wolves. Remind students that comparing sentences tell how things are alike and contrasting sentences tell how things are different. Tell students that effective sentences flow smoothly from one to the next. Explain that when all of the sentences in an essay are of the same structure and length, the essay may be boring. Remind them to use a variety of sentence lengths. Tell students to support their facts with specific details.

Fluency

Oral Reading Fluency

CM 117

Assessment Use Oral Reading Fluency Practice page 117 to assess students' oral reading fluency.

Name ______________________ **Wolf Tales** Vocabulary

▶ **Fill in the chart with the word or words that mean the opposite of each word.**

tender, delighted, dusk, brittle, cunning, latch, embraced

Vocabulary Word	Antonym
delighted	**1.** sad
dusk	**2.** dawn
embraced	**3.** push away
latch	**4.** unlock
tender	**5.** rough
brittle	**6.** strong
cunning	**7.** innocent

▶ **Write a paragraph about a time when you felt delighted. Use three vocabulary words.**

Possible response: I was delighted when my baby sister came home from the hospital. I embraced her and gave her a tender kiss.

▲ Vocabulary Practice p. 115

Name ______________________ **Wolf Tales** Comprehension

▶ **Circle the best answer to each question below.**

1. Which would NOT be included in a summary of *Peter and the Wolf*?
- **A** Peter tricked Great-Uncle Wolf into releasing a bird.
- **B** Great-Uncle Wolf ended up living in a zoo.
- Ⓒ Peter roamed too far from his home.
- **D** Great-Uncle Wolf swallowed a duck.

2. Which is a similarity of all three tales?
- **F** A character wears a costume.
- Ⓖ A character plays a trick on someone.
- **H** A character is never heard from again.
- **J** A character tries to eat another animal.

3. Which event happened first?
- **A** Great-Grandma Wolf sewed a sheep's costume.
- **B** A shepherd grew hungry watching his sheep.
- Ⓒ Great-Grandma Wolf wanted lamb chops.
- **D** A shepherd was fooled by a wolf.

▶ **Use the diagram below to compare and contrast the wolves from *A Wolf in Sheep's Clothing* and *Little Red Riding Hood*.**

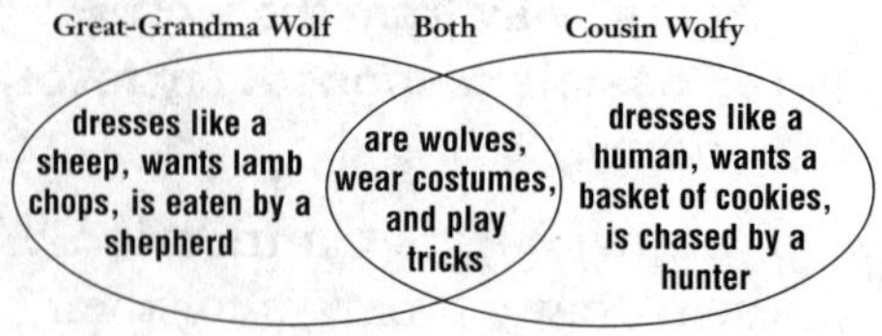

▲ Comprehension Practice p. 116

Name ________________________________

▶ **Fill in the chart with the word or words that mean the opposite of each word.**

tender | **delighted** | **dusk** | **brittle**
cunning | **latch** | **embraced**

Vocabulary Word	Antonym
	1. sad
	2. dawn
	3. push away
	4. unlock
	5. rough
	6. strong
	7. innocent

▶ **Write a paragraph about a time when you felt delighted. Use three vocabulary words.**

____

____

____

____

____

Name ____

▶ **Circle the best answer to each question below.**

1. Which would NOT be included in a summary of *Peter and the Wolf*?

A Peter tricked Great-Uncle Wolf into releasing a bird.
B Great-Uncle Wolf ended up living in a zoo.
C Peter roamed too far from his home.
D Great-Uncle Wolf swallowed a duck.

2. Which is a similarity of all three tales?

F A character wears a costume.
G A character plays a trick on someone.
H A character is never heard from again.
J A character tries to eat another animal.

3. Which event happened first?

A Great-Grandma Wolf sewed a sheep's costume.
B A shepherd grew hungry watching his sheep.
C Great-Grandma Wolf wanted lamb chops.
D A shepherd was fooled by a wolf.

▶ **Use the diagram below to compare and contrast the wolves from *A Wolf in Sheep's Clothing* and *Little Red Riding Hood*.**

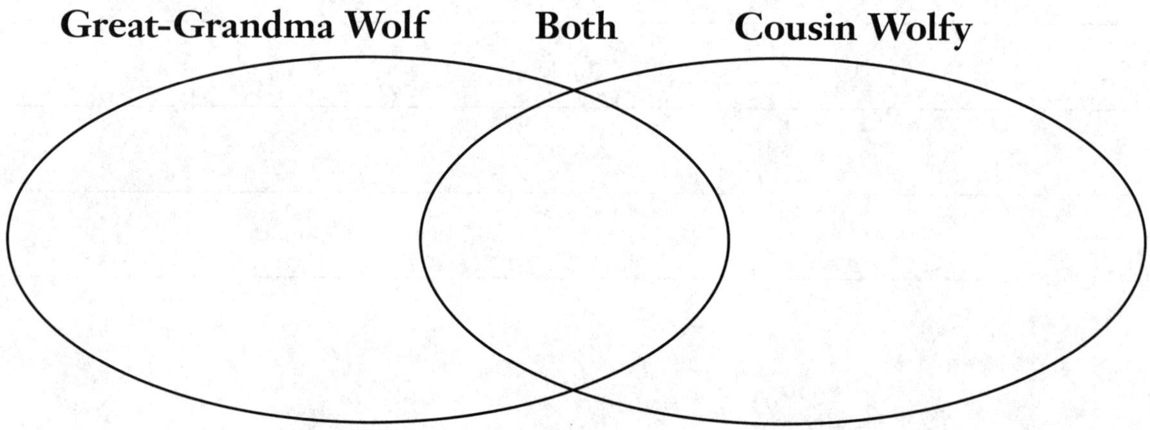

Name ______________________________

"That's right," Papa Wolf replied proudly. "The forest is where many 11
members of our family have lived. Your Great-Uncle Wolf is the wolf who 24
met Peter." 26

"How did it happen?" the third pup asked. 34

"Well," Papa Wolf explained, "Peter had been warned not to roam too far 47
from home. He was told that the wolf might come near." 58

"What's wrong with that?" the first pup asked. "Isn't Great-Uncle Wolf 69
friendly?" 70

"Yes, he is," Papa Wolf sighed. "But, sadly, not all humans understand 82
that." 83

"So, what happened when he met Peter?" the second pup asked. 94

Papa Wolf continued. "Peter was outdoors, playing with the ducks and 105
birds and frogs. It was dusk, just about dinner time. Along came your Great- 119
Uncle Wolf, who was quite hungry. He spotted a tender young duck and— 132
GULP!—had a fast-food meal. He swallowed the whole duck in one bite." 145

"Wow!" the first pup gasped. 150

"Did Peter get upset?" the second pup asked. 158

"Of course," Papa Wolf replied. 163

FLUENCY SCORE

Total Words Read Per Minute ____________

Number of Errors — ____________

Number of Words Read Correctly ____________

Trying Out

by Kate Pistone

SUMMARY Some friends try out for parts in a play about Pocahontas and John Smith. Although some expect to get good roles in the play, it turns out that those who practice hard and learn all their lines get the lead roles.

- **Focus Skill:** Author's Purpose
- **Focus Strategy:** Use Text Structure and Format
- **Vocabulary:** wits, wailing, faring, advice, dreadful, farewell

Differentiated Instruction

Paired Reading Organize students in pairs. Have them read the selection aloud to each other, alternating with each page. After students have finished reading, have them discuss in pairs a part of the play they found interesting.

Prepare to Read

Preview/Set Purpose Have students preview the play by reading the table of contents, stage directions, and characters' names. Ask them what they think they will learn from this play.

Focus Skill

Author's Purpose

Teach/Model Tell students that authors have different purposes, or reasons, for writing. Review the three main purposes: to inform, to persuade, and to entertain. Remind students that authors can have more than one purpose for writing. Display a chart similar to the one below. Ask them to use the chart to decide what the author's purpose was for writing each quote.

Quotation from Play	Author's Purpose
"I'm sure to get the part of John Smith. He was the leader of the Jamestown Colony."	To inform
"Cut out the talking. Pocahontas probably knew when to keep quiet."	To entertain

Practice/Apply Have students complete the chart by finding one more quotation that informs and one more quotation that entertains.

Focus Strategy

Use Text Structure and Format

Teach/Model Remind students that paying attention to text structure and format can help them better understand what they are reading. Tell students that *Trying Out* is a play, and that plays have special formats. Plays are divided into scenes. At the beginning of each scene, the author describes the time, the setting, and what the characters are doing. The characters' names are printed in capital letters and are followed by their dialogue. Instructions that tell the characters what to do are printed in italics. Model the strategy by directing students' attention to page 2. Say, *I know from the text structure and format that this is Scene I of the play. I know that the setting is the first day of Adams Acting Camp. I know the characters are Lauren, Kari, Max, Andy, Paul, Sara, and Mrs. Adams.*

Practice/Apply Have students find two examples of stage directions as they read. Ask them to explain how the stage directions help them understand the play.

Build Vocabulary

Help students use context to understand key vocabulary. Have them answer each question by reading the words and sentences that surround the word.

wits (p. 5)	Why do you need your *wits* to do well in school? **Wits help you think.**
wailing (p. 11)	If someone was *wailing*, what would they be doing? **crying loudly**
faring (p. 6)	What word or phrase can you use in place of *faring*? **doing**
advice (p. 6)	Who are good people to ask for *advice*? **parents, teachers, friends**
dreadful (p. 6)	What word means the opposite of *dreadful*? **good, wonderful, excellent**
farewell (p. [illegible])	When should you say *farewell* to someone? **right before you leave**

CM 121 **Independent Practice:** Vocabulary Practice page 121

Read and Comprehend

Monitor Comprehension

Guide students as they read the text, using the following questions:

Pages

2–4 How does the author inform readers about Pocahontas? **Lauren says that Pocahontas was smart, beautiful, and saved Captain Smith's life twice.** AUTHOR'S PURPOSE (Focus Skill)

5–7 How does Andy feel about trying out for the play? **He is confident because he has experience and he thinks he is good.** CHARACTER'S EMOTIONS

Who does Max ask for advice? **Max asks Paul and Sara, the counselors.** NOTE DETAILS

8–9 When does Scene II take place? **It takes place a week after auditions.** LOCATE INFORMATION

10–12 How does Lauren react to the cast list? **She cries because she does not get the role of Pocahontas.** CHARACTER'S EMOTIONS

13–16 What lesson is the author trying to teach the reader? **Hard work pays off.** AUTHOR'S PURPOSE (Focus Skill)

How does the text structure help you understand Andy's reaction to the cast list? **Information in italics reveals Andy's feelings.** TEXT STRUCTURE AND FORMAT (Focus Strategy)

Think and Respond

Have students answer the questions on the inside back cover of the book.

1. **Possible response: It is unusual because it is written in play form, with scenes instead of chapters.**
2. **Max worked hard and Andy did not. The writer has Max get a part to show that it pays to work for what you want.**
3. **It is titled "Trying Out" because the kids are trying out for the play.**
4. **Possible response: I learned that John Smith was the leader of the colony.**
5. **Lauren and Kari are alike because they both want the part of Pocahontas. They are different because Kari practiced a lot for the part, but Lauren did not because she had soccer practice and piano lessons.**
6. **Possible response: I liked Max the best because he overcame his fear and got a part in the play.**

SOCIAL STUDIES

Social Studies Connection Have students research how colonists and Native Americans dressed during the 1600s in Jamestown.

School-Home Connection Tell students to find out about Native American food. Have them pick their favorite recipe and ask an adult to help them prepare it at home.

CM 122 **Independent Practice:** Comprehension Practice page 122

Write

Writer's Craft: Word Choice

Write a Paragraph that Explains Have students read through the dialogue in *Trying Out* and note the adjectives. Say, *By using adjectives, the writer helps us to understand what the characters are doing. On the first page, the writer uses the words* first *and* pleased *to describe something or someone.* Tell students that good writers use adjectives to describe people, places and things in a story. Tell them to write a paragraph using adjectives to describe what they would do in a play. Remind them to use varying sentence lengths and interesting details as they write.

Fluency

Oral Reading Fluency

CM 123

Assessment Use Oral Reading Fluency Practice page 123 to assess students' oral reading fluency.

Name ______________________ Trying Out Vocabulary

▶ **Write the word that best completes each sentence.**

wits wailing advice dreadful faring farewell

1. The baby's **wailing** woke up her parents.
2. When I have a problem, I usually ask for **advice**.
3. I say **farewell** to my teacher at the end of the day.
4. To solve a puzzle you need to use your **wits**.
5. I went home early because I was feeling **dreadful**.
6. My friend is **faring** well at his new school.

▶ **Replace the underlined word(s) with the best vocabulary word.**

7. Luis was in a terrible mood because he had a bad day. **dreadful**
8. Sometimes, following other people's ideas on how to solve a problem is a mistake. **advice**
9. The loud cries of the child could be heard throughout the neighborhood. **wailing**
10. Mrs. Parker gave me a sticker because I was doing so well in school! **faring**

▲ Vocabulary Practice p. 121

Name ______________________ Trying Out Comprehension

▶ **Answer the following questions about *Trying Out*.**

1. What is the format of *Trying Out*?
   It is a play.
2. How is the play organized?
   It is divided into three scenes.
3. What is Sara's advice to Max?
   to do relaxing exercises so he won't be nervous
4. Why didn't Andy practice for the audition?
   He thought he didn't need to because he has experience.

▶ **Write a letter to Andy persuading him to practice for the audition.**

Possible response: Dear Andy, You are trying out for the lead role, so this play must be important to you. Even if you have experience, you still need to practice to make sure you do your best. If you don't know your lines, you won't get the part you want.

▲ Comprehension Practice p. 122

Name ____________________

Write the word that best completes each sentence.

wits	wailing	advice	dreadful	faring	farewell

1. The baby's ____________ woke up her parents.
2. When I have a problem, I usually ask for ____________.
3. I say ____________ to my teacher at the end of the day.
4. To solve a puzzle you need to use your ____________.
5. I went home early because I was feeling ____________.
6. My friend is ____________ well at his new school.

Replace the underlined word(s) with the best vocabulary word.

7. Luis was in a <u>terrible</u> mood because he had a bad day.

   ____

8. Sometimes, following other people's <u>ideas on how to solve a problem</u> is a mistake. ____________
9. The <u>loud cries</u> of the child could be heard throughout the neighborhood. ____________
10. Mrs. Parker gave me a sticker because I was <u>doing</u> so well in school!

    ____

Name ______________________________

▶ **Answer the following questions about *Trying Out*.**

1. What is the format of *Trying Out*?

____

2. How is the play organized?

____

3. What is Sara's advice to Max?

____

4. Why didn't Andy practice for the audition?

____

▶ **Write a letter to Andy persuading him to practice for the audition.**

____

____

____

____

Name ______________________________

MRS. ADAMS: I hope all of you are enjoying your first day at Adams 14
Acting Camp. I am pleased that all of you could make it. 26
LAUREN: I can't wait to hear about the play! 35
MRS. ADAMS: This year's play is about the first settlers of the Jamestown 48
Colony and their Native American neighbors. The main characters are 58
Pocahontas, her father the great Chief Powhatan, and Captain John Smith. 69
The setting is in Virginia in the early 1600s. Some of you will be members 84
of Powhatan's tribe. Some of you will be colonists. 93
LAUREN: I want to be Pocahontas! She saved Captain Smith's life twice. 105
She was smart, and I think she was beautiful. You should see the portrait 119
made of her when she went to England and . . . 128
PAUL: Shh! Please be quiet. Everyone is trying to listen to the director. 141
ANDY: Yeah, cut out the talking. Pocahontas probably knew when to keep 153
quiet. 154
MRS. ADAMS: I know you're all anxious to know what parts you will 167
have. 168

FLUENCY SCORE
Total Words Read Per Minute ____________
Number of Errors — ____________
Number of Words Read Correctly ____________

Student of the Week

by Stef Schumacher

SUMMARY: During the Depression, Thomas reminds his family and unemployed father that "hope is free." When his father finds a job, Thomas writes an account of their suspenseful week and wins the Student of the Week award he has longed for.

- **Focus Skill:** Fact and Opinion
- **Focus Strategy:** Self-Question
- **Vocabulary:** county, auctioneer, bid, clutched, glistened, galloped

Differentiated Instruction

Independent Reading Have students read the story independently. Encourage students to write down questions to ask if they do not understand a part of the story.

Prepare to Read

Preview/Set Purpose Have students preview the book by looking at the illustrations. Point out that this is a realistic fiction book, which is a fictional story about something that could happen. Ask them what they think they will learn from this book. Have students set a purpose for reading, such as *I want to find out who becomes student of the week.*

Fact and Opinion

Teach/Model Explain to students that facts are events that really occurred. They can be proven. Tell students that opinions are a person's thoughts or feelings. Have students read page 2 of the story. Say, *The statement, "Every week, his teacher, Mrs. Johnson, had the class write a story" is a fact because it can be proven. But the last sentence on the page, "Times were tough, and his family needed his help" is an opinion because it explains what the author feels.*

Practice/Apply Have students read pages 4–5 of the book and find a fact and an opinion.

Self-Question

Teach/Model Tell students that good readers self-question to understand important ideas and solve problems as they read. Point out that when good readers come to a part of a story they find confusing, they ask themselves a question to try to clear up the confusion. Have students read page 2. Say, *When I read page 2, I ask myself: Why does Thomas study before school? That question is answered right away: he works at the grocery store in the afternoon. Why does Thomas work at the grocery store? I guess I'll find out from the rest of the story.*

Practice/Apply Have students self-question whenever they find something confusing in the story.

Build Vocabulary

Help students use context to understand key vocabulary. Have them answer each question by reading the words and sentences that surround the word.

county (p. 9)	What's bigger: a *county* or a country? **a country**
auctioneer (p. 7)	Would an *auctioneer* work in a grocery store? **No, you don't bid in a store.**
bid (p. 6)	Where do you *bid* for something? **Possible response: at an auction**
clutched (p. 14)	Why would you *clutch* a hundred dollars? **Possible response: so I wouldn't drop it**
glistened (p. 9)	What might be a synonym for *glistened*? **shimmered, shone**
galloped (p. 14)	How did the horse move if it *galloped*? **It moved quickly.**

CM 127 **Independent Practice:** Vocabulary Practice page 127

Read and Comprehend

Monitor Comprehension

Guide students as they read the text, using the following questions:

Pages

2–3 Why does Thomas work at Mr. Barton's grocery story? **His family needs his help.** CHARACTER'S MOTIVATIONS

4–7 What fact makes Dad think he shouldn't try to get the job at the shipyard? **He is a carpenter.** FACT AND OPINION (Focus Skill)

8–10 Why does Thomas smile when he hears his dad leave the house? **He is glad that his dad is trying to get the job.** CHARACTER'S EMOTIONS

Why is it hard for Thomas to pay attention in class? **He is worried about his family.** MAKE INFERENCES

11–13 What is Dad's opinion about what will happen? **He thinks things will get better.** FACT AND OPINION (Focus Skill)

14–16 Do you think that Mr. Barton is a kind man? **Yes. He gives Thomas and his family groceries for free.** EXPRESS PERSONAL OPINION

What is Thomas's story about? **Possible response: his family's struggle while Dad was out of work** SUMMARIZE

Think and Respond

Have students answer the questions on the inside back cover of the book.

1. **This story takes place in 1931.**
2. **The teacher's statement is an opinion.**
3. **Thomas is a hard worker in school, at his job, and at home. He helps his family by working after school at the grocery store to make money while his father is out of work.**
4. **Possible response: I learned that times were hard and it was difficult to find a job.**
5. **The two times are the same because he is up against other people for the job. They are different because one is an auction setting, and the other is not.**
6. **Possible response: I would tell a story of the difficulties my family has faced and how we overcame them.**

SOCIAL STUDIES **Social Studies Connection** Have students work in groups to brainstorm what they know about the Great Depression. See if they know the causes, how long it lasted, and how people were affected.

School-Home Connection Have students imagine they are grandparents, and write a paragraph describing their experiences to their grandchildren.

CM 128 **Independent Practice:** Comprehension Practice page 128

Write

Writer's Craft: Organization

Write a Research Report Have students read the first paragraph on page 5 of *Student of the Week*. Tell them that the author organized the paragraph by starting with a main idea and supporting it with details. Explain that good writers organize information so that readers will understand it. Tell students that when writing a research report it is important to organize every paragraph. The best way to organize is to outline, or prewrite, the information you will present. Have students research and write a paragraph about a grocery store and the different jobs people have there.

Fluency

Oral Reading Fluency

CM 129

Assessment Use Oral Reading Fluency Practice page 129 to assess students' oral reading fluency.

Name ____________________

Student of the Week
Vocabulary

▶ **Write the word that best completes the sentence.**

county auctioneer bid clutched glistened galloped

1. Dad put in a **bid** for the old red wagon.
2. Paul Revere **galloped** through Boston on his horse.
3. When we went hiking, I **clutched** my sister's hand so I wouldn't fall.
4. My grandparents live in a different **county**, but we still live in the same state.
5. The stars **glistened** in the sky as we built a campfire.
6. I tried to win the new bike at the sale, but the **auctioneer** didn't see me raise my hand.

▶ **Imagine you are a newspaper reporter at an auction. Write a paragraph about the event. Use at least three of the vocabulary words.**

Possible response: Today there was a job auction where eyes glistened with hope at the thought of a job. People came from all over the county. Many people clutched their chairs as the jobs were announced. Two lucky people got jobs, and many walked out as they came in—unemployed.

Name ____________________

Student of the Week
Comprehension

▶ **The chart below lists some questions and answers about *Student of the Week*. Fill in the missing questions and answers.**

Questions I Ask Myself	Answers
What will Thomas write about?	**how his family got through tough times**
What is Thomas's job at the store?	He delivers groceries and keeps the store neat and clean.
What will happen at the auction?	Two other people get the jobs.
Will Thomas have to give up anything to help his family?	**No, he can stay in school.**

▶ **Write F next to facts from the story, and O next to opinions.**

F 1. Thomas studies before school.
O 2. Winning the Student of the Week award would be an exciting honor.
F 3. Thomas's father does not have a job when the story begins.
F 4. Mr. Grant and Mrs. Mathews sell things to make money for their families.
O 5. Thomas is one of Mr. Barton's best workers.
F 6. Dad decides to go to the auction.
O 7. It is terrible that Thomas's father does not get the job at the shipyard.
F 8. Thomas hears his parents talking while he is in bed.
F 9. Thomas makes a chart of the things he has to do for the day.
O 10. It is very nice of Paul to do Thomas's chores.

▲ Vocabulary Practice p. 127

▲ Comprehension Practice p. 128

Name ______________________________

▶ **Write the word that best completes the sentence.**

county	**auctioneer**	**bid**	**clutched**	**glistened**	**galloped**

1. Dad put in a ______________ for the old red wagon.
2. Paul Revere ______________ through Boston on his horse.
3. When we went hiking, I ______________ my sister's hand so I wouldn't fall.
4. My grandparents live in a different ______________, but we still live in the same state.
5. The stars ______________ in the sky as we built a campfire.
6. I tried to win the new bike at the sale, but the ______________ didn't see me raise my hand.

▶ **Imagine you are a newspaper reporter at an auction. Write a paragraph about the event. Use at least three of the vocabulary words.**

____

____

____

____

____

Name ______________________________

▶ **The chart below lists some questions and answers about *Student of the Week*. Fill in the missing questions and answers.**

Questions I Ask Myself	Answers
What will Thomas write about?	
	He delivers groceries and keeps the store neat and clean.
	Two other people get the jobs.
Will Thomas have to give up anything to help his family?	

▶ **Write F next to facts from the story, and O next to opinions.**

____ **1.** Thomas studies before school.

____ **2.** Winning the Student of the Week award would be an exciting honor.

____ **3.** Thomas's father does not have a job when the story begins.

____ **4.** Mr. Grant and Mrs. Mathews sell things to make money for their families.

____ **5.** Thomas is one of Mr. Barton's best workers.

____ **6.** Dad decides to go to the auction.

____ **7.** It is terrible that Thomas's father does not get the job at the shipyard.

____ **8.** Thomas hears his parents talking while he is in bed.

____ **9.** Thomas makes a chart of the things he has to do for the day.

____ **10.** It is very nice of Paul to do Thomas's chores.

Name ______________________________

Student of the Week
Oral Fluency

Thomas delivered groceries for Mr. Barton. As he pushed 9
the cart through the city, he noticed that people all around him 21
were struggling to make a living. Mr. Grant sold apples for a 33
nickel apiece. Mrs. Mathews sold old furniture, books, and 42
clothes to make money for her family. 49

Thomas felt lucky to have his job at the grocery store. Some 61
days he wished he could go home and study after school, but 73
his family needed the money he earned to help pay the rent and 86
to buy food. 89

Back at Mr. Barton's grocery store, Thomas straightened all 98
the items on the shelves and swept the floors. 107

"I like to help my best workers," Mr. Barton said. "There's 118
going to be an auction tomorrow for a new job over at the 131
shipyard. Tell your dad to go and put in a bid first thing in the 146
morning." 147

"I will, Mr. Barton! I will!" shouted Thomas. 155

Thomas ran all the way home, happy that he had good news. 167

FLUENCY SCORE

Total Words Read Per Minute ____________

Number of Errors – ____________

Number of Words Read Correctly ____________

Tall Tales

Pecos Bill

by Michael Burgan

SUMMARY Real cowhands enjoyed telling tall tales about the incredible exploits of Pecos Bill. This book recounts some of these stories, including his ride on a tornado and his creation of the Grand Canyon.

- **Focus Skill:** Main Idea and Details
- **Focus Strategy:** Use Context to Confirm Meaning
- **Vocabulary:** ranchers, tending, profit, corral, stray, market

Differentiated Instruction

Independent Reading Have students read through the selection independently. Encourage students to write down questions to ask if they do not understand a part of the story.

Prepare to Read

Preview/Set Purpose Have students preview the book by looking at the illustrations. Ask students what they think the book will be about. Have students set a purpose for reading, such as *I want to know the stories about Pecos Bill.*

Focus Skill

Main Idea and Details

Teach/Model Display a chart similar to the one shown below. Tell students that authors sometimes organize selections by main idea and details to make the information easy to understand. Explain that the main idea is what a selection or part of a selection is mostly about and that details help explain the main idea. Have students read page 2 of the story. Say, *The main idea of this paragraph is that Pecos Bill was the best rider in the Old West. "Bill rode a mountain lion" is a detail that gives more information about him.*

Main Idea	Details
Pecos Bill was the best rider in the Old West.	*Bill rode a mountain lion.*

Practice/Apply Have students read page 5 of the story. Have them create their own charts like the one above and write the main idea and details in it.

Focus Strategy

Use Context to Confirm Meaning

Teach/Model Tell students that good readers use context to confirm meaning as they read. If they come to an unfamiliar word, they use the words and phrases around the word to confirm its meaning. Have students read page 4. Say, *I'm not sure what the word* legend *means. So I read the sentences around it and learn that the Pecos Bill stories are set in the West, but they're not supposed to be believed. Since the author also uses tall tales, which are exaggerated stories, as a synonym for* legend, *I think* legend *means* a story from long ago.

Practice/Apply Have students use context to confirm the meanings of unfamiliar words.

Build Vocabulary

Help students use context to understand key vocabulary. Have them answer each question by reading the words and sentences that surround the word.

ranchers (p. 10)	What do *ranchers* do? **They raise farm animals, like cows or horses.**
tending (p. 4)	What does *tending* mean? **taking care of**
profit (p. 11)	Is a *profit* the same as a loss? **No, a profit is money gained from a sale.**
corral (p. 10)	How does a *corral* hold farm animals? **with a fence**
stray (p. 2)	Why do animal shelters care for *stray* animals? **because they have no home**
market (p. 11)	Describe a *market*. **People buy and sell things like fruits and vegetables**

CM 133 **Independent Practice:** Vocabulary Practice page 133

Read and Comprehend

Monitor Comprehension

Guide students as they read the text, using the following questions:

Pages

2–5 Who invented Pecos Bill? **No one is sure. Some think cowboys and cowgirls, but Edward O'Reilly said he invented Pecos Bill.** MAIN IDEA AND DETAILS

6–8 If you didn't know what a coyote was, what details would help you find out? **animals with tails that howl at the moon; also, the illustration** USE CONTEXT TO CONFIRM MEANING

How did Bill keep busy? **He went after all the robbers in Texas.** PROBLEM-SOLUTION

9–11 Where was Bill's ranch? **It started in Texas and spread all over New Mexico.** SETTING

12–13 What happened when the tornado cut into rocks? **It created the Grand Canyon.** CAUSE-EFFECT

14–16 How do the pictures show that the story is a tall tale? **The events shown in the pictures could not actually happen.** USE GRAPHIC AIDS

Think and Respond

Have students answer the questions on the inside back cover of the book.

1. **The story called "Young Pecos Bill" tells about him as a child.**
2. **The writer says that the stories about Pecos Bill are not true, that he was not a real cowboy, and he never lived in the Old West. The writer also says that the stories are legends or tall tales.**
3. **Possible response: People enjoy them because they are humorous and fun to read.**
4. **Possible response: I learned what states were in the Old West.**
5. **The two meetings were similar because both the rattlesnake and the mountain lion gave up in their fights against Pecos Bill, and Pecos Bill kept them both as pets. They were different because the snake gave up sooner than the mountain lion, who wrestled Pecos Bill for hours.**
6. **Possible response: I liked the tale about Pecos Bill and Slue Foot Sue because I thought it was funny when she landed on her puffy dress and started bouncing.**

Science Connection Have students look up facts about twisters, the Grand Canyon, or Death Valley. Have them compare it to the Pecos Bill stories.

School-Home Connection Have students work with a friend or relative to list the qualities of a tall tale.

CM 134 **Independent Practice:** Comprehension Practice page 134

Write

Writer's Craft: Organization

Write a Research Report Have students read the last paragraph on page 3. Say *The main idea is that people are not supposed to believe the tales about Pecos Bill. One detail would be that these stories take place during the American West.* Tell students that an outline helps to organize facts and information. Explain that an outline uses an alphabetical or numerical system to put main ideas and supporting details in order. An outline should start with an introduction. Then it should list subsections that will make up the body of a report. Finally, an outline should have a conclusion. Have students write an outline for a research report on Pecos Bill.

Fluency

Oral Reading Fluency

CM 135

Assessment Use Oral Reading Fluency Practice page 135 to assess students' oral reading fluency.

Name ____________________ **Pecos Bill Vocabulary**

▶ **Read the paragraph below. Then fill in the word that fits in the blank.**

ranchers tending profit corral stray market

My Uncle Robert owns a ranch in Montana. He keeps horses and cows in a **corral**. Last summer, I went to visit the ranch. I helped him by **tending** the cows. We spent a whole afternoon looking for a **stray** calf. It was hard work, but Uncle Robert said I was a great cowhand! Uncle Robert also took me to the **market** one day. I met many **ranchers** there. Most of them said they were going to make **profit** that year.

▶ **Describe what you might see if you visited a ranch. Use at least three vocabulary words.**

Possible response: Several ranchers are tending to the animals. The ranchers gather all the stray cows and lead them into a corral. These animals may bring a large profit to the ranchers.

Books for All Learners Lessons, Grade 3 133

▲ Vocabulary Practice p. 133

Name ____________________ **Pecos Bill Comprehension**

▶ **Circle the best answer for each question below.**

1. Choose the sentence that supports this main idea: Pecos Bill knew that riding a twister would be easy.
- **A** People thought he could do anything.
- **B** A cowboy dared him to ride a tornado.
- **C** The tornado destroyed the countryside around him.
- **(D)** He went and jumped on a big tornado.

2. Which of the following is a context clue that helps you confirm the meaning of *pounced*?
- **F** Pecos Bill put a saddle on the mountain lion and rode it.
- **G** The mountain lion got tired, and gave up, like the snake.
- **(H)** Pecos Bill saw the mountain lion sitting up on a cliff.
- **J** The cat was as big as ten bulls and a whole lot meaner.

3. What is the most important idea in *Pecos Bill*?
- **A** Cowhands started talking about Pecos Bill years ago.
- **(B)** The Pecos Bill stories are entertaining, but not true.
- **C** Pecos Bill was a real cowboy and an American hero.
- **D** The Pecos Bill stories are based on real events.

▶ **What parts of the Pecos Bill stories confirm that they are tall tales?**

Possible response: his impossible tasks such as riding tornadoes, roping the Rio Grande, and wrestling mountain lions; the large size of everything such as Sue's catfish, the lion, and Bill's ranch

134 Books for All Learners Lessons, Grade 3

▲ Comprehension Practice p. 134

Name ______________________________

▶ **Read the paragraph below. Then fill in the word that fits in the blank.**

ranchers **tending** **profit** **corral** **stray** **market**

My Uncle Robert owns a ranch in Montana. He keeps horses and cows in a ______________. Last summer, I went to visit the ranch. I helped him by ______________ the cows. We spent a whole afternoon looking for a ______________ calf. It was hard work, but Uncle Robert said I was a great cowhand! Uncle Robert also took me to the ______________ one day. I met many ______________ there. Most of them said they were going to make ______________ that year.

▶ **Describe what you might see if you visited a ranch. Use at least three vocabulary words.**

____

____

____

____

____

____

Name ______________________________

▶ Circle the best answer for each question below.

1. Choose the sentence that supports this main idea: Pecos Bill knew that riding a twister would be easy.
   A People thought he could do anything.
   B A cowboy dared him to ride a tornado.
   C The tornado destroyed the countryside around him.
   D He went and jumped on a big tornado.

2. Which of the following is a context clue that helps you confirm the meaning of *pounced*?
   F Pecos Bill put a saddle on the mountain lion and rode it.
   G The mountain lion got tired, and gave up, like the snake.
   H Pecos Bill saw the mountain lion sitting up on a cliff.
   J The cat was as big as ten bulls and a whole lot meaner.

3. What is the most important idea in *Pecos Bill*?
   A Cowhands started talking about Pecos Bill years ago.
   B The Pecos Bill stories are entertaining, but not true.
   C Pecos Bill was a real cowboy and an American hero.
   D The Pecos Bill stories are based on real events.

▶ What parts of the Pecos Bill stories confirm that they are tall tales?

____

____

____

____

Name ____

Bill had heard of a cowgirl named Slue Foot Sue. Folks said she 13
could ride almost as well as Bill himself. When Bill met Sue, she 26
was riding a catfish. The fish was as big as a whale. Sue rode the 41
catfish up and down the Rio Grande. Bill was impressed. He asked 53
Sue to be his wife. 58

On her wedding day, Sue wore a long, beautiful dress. After the 70
wedding, Bill said to her, "Now Sue, I have just one favor to ask 84
you. Please don't ever ride Widow Maker." 91

Widow Maker was Bill's horse. She had been wild until Bill 102
tamed her. Only he could safely ride Widow Maker. Sue wanted to 114
please Bill, but she wanted to ride Widow Maker, too. Before Bill 126
could stop her, she jumped on the horse. Widow Maker bucked and 138
tossed Sue into the sky. 143

She sailed so high, she almost hit the moon! Then she fell down 156
to the ground and landed on her puffy dress. 165

FLUENCY SCORE

Total Words Read Per Minute ____________

Number of Errors – ____________

Number of Words Read Correctly ____________

Realistic Fiction

A Trip to the Past

by Laurie Rozakis

SUMMARY A modern family visits the preserved historical town of Leadville, Colorado, where their guidebook and a museum docent enlighten them about miners' lives during the Silver Rush of the 1860s.

- **Focus Skill:** Fact and Opinion
- **Focus Strategy:** Self-Question
- **Vocabulary:** stagecoach, miners, nuggets, skillet, settled, boom town, landmark

Differentiated Instruction

Independent Reading Have students read the selection independently. Have students write down questions to ask if they do not understand a part of the selection.

Prepare to Read

Preview/Set Purpose Have students examine the cover and title. Have students preview the book by looking at the illustrations and signs. Have them explain how these will be helpful to them. Point out that this book is realistic fiction, which is a made-up story about something that could happen. Ask them what they think they will learn from this book.

Focus Skill

Fact and Opinion

Teach/Model Display a fact and opinion chart similar to the one below. Tell students that good readers are alert to which statements can be proven and which tell what someone thinks. Fill in the chart based on pages 2–3. Ask students to give an example of a fact and of an opinion.

Fact	Opinion
The old mine is at the end of the street. **Many people came to Leadville after silver was found there**	Visitors should go to the snack bar first. **Leadville is an interesting place to visit.**

Practice/Apply Have students add to their fact and opinion charts after they read *A Trip to the Past.*

Focus Strategy

Self-Question

Teach/Model Tell students that good readers use the Self-Questioning strategy to help them understand important ideas and solve problems as they read. Point out that when good readers get confused, they ask a question. Have students read page 2. Say, *When I read, I ask myself: Why would people keep the city looking as it did in the 1870s? I'll have to keep reading for the answer. What will the Kims do first? That question is answered right away. They look at the map.* Point out that self-questioning about fact and opinion can also help readers gain a better understanding of what is happening and what the characters think of it.

Practice/Apply Have students use the Self-Questioning strategy whenever they read something confusing.

Build Vocabulary

Help students use context to understand key vocabulary. Have them answer each question by reading the words and sentences that surround the word.

stagecoach (p. 4)	Instead of a *stagecoach*, what do we use? **Possible answer: cars**
miners (p. 2)	What do *miners* look for? **Possible response: silver, gold, and other metals**
nuggets (p. 14)	What could a miner buy with silver *nuggets*? **Possible response: food**
skillet (p. 9)	What can you cook in a *skillet*? **Possible responses: eggs, bacon**
settled (p. 4)	Why would someone have *settled* in Florida? **Possible response: The weather is warm.**
boom town (p. 6)	How do *boom towns* grow? **quickly**
landmark (p. 3)	Name a U. S. *landmark*. **Possible response: the White House**

CM 139 **Independent Practice:** Vocabulary Practice page 139

Read and Comprehend

Monitor Comprehension

Guide students as they read the text, using the following questions:

Pages

2–4 What changed in Leadville in 1877? **Silver was found; people moved in.** STORY EVENTS

5–8 What fact does Mrs. Kim tell about Leadville in 1880? **It is the world's largest source of silver.** FACT AND OPINION

9–10 Ask yourself what the biggest danger was that the miners faced. **rats that spread sickness** SELF-QUESTIONING

How did miners waterproof their boots? **with bear fat** CAUSE-EFFECT

11–13 What is Mrs. Kim's opinion about the miners? **Their lives must have been hard.** FACT AND OPINION

14–16 Do you think the stories of the giant nuggets were true? Explain. **Possible response: No, he's just trying to capture the interest of the kids.** FACT AND FANTASY

Think and Respond

Have students answer the questions on the inside back cover of the book.

1. **Possible response: The glossary helps me read the story by letting me know meanings of words that I have not heard before.**
2. **The facts in the story are the parts that explain what a boom town is and why Leadville is famous. The opinions in the story are the parts where Tracy and Tod comment on what they think of how miners lived.**
3. **Possible response: Historical museums are important because visiting a museum is an interesting way to learn about history by seeing real objects from the past.**
4. **Possible response: I learned that a miner's life in the 1870s was difficult. The miners often were hungry and dirty, and they worked hard for very little money.**
5. **Leadville was a small town of 200 people in 1877, but it grew to over 15,000 people in three years' time. The people there searched for silver and the town had a general store, a hotel, and a dance hall. Today, Leadville looks the same as it did back then, but miners no longer live there. Visitors come to see the past.**
6. **Possible response: I would want to see the museum in Leadville because it would be interesting to see things from the 1800s.**

SOCIAL STUDIES

Social Studies Connection Have students pretend they are miners in Leadville and write letters to their families.

School-Home Connection Have students draw their town as they imagine it looked one hundred years ago.

CM 140 **Independent Practice:** Comprehension Practice page 140

Write

Writer's Craft: Organization

Write a Research Report Tell students that the draft of a research report should organize information so it is easy for the audience to understand. Explain that an introduction states the main idea and should be a good "hook" to grab the audience's attention. Details and examples are added to the introduction to support the main idea. Remind students that writing an outline can help writers list the details they want to include. Work with students to write a paragraph about a boom town. Remind them to include an introduction that expresses the main idea in an interesting way and to add detail sentences that support the main idea, using an outline if necessary.

Fluency

Oral Reading Fluency

CM 141

Assessment Use Oral Reading Fluency Practice page 141 to assess students' oral reading fluency.

Name ______________________

A Trip to the Past
Vocabulary

▶ **Write the word that best completes the sentence.**

miners	landmark	settled	stagecoach	nuggets
boom town	skillet			

1. **Miners** left Leadville when the town ran out of silver.
2. People traveled for miles and **settled** in Leadville.
3. Mrs. Kim sees on a map that the General Store is the town's **landmark**.
4. One way miners traveled to Leadville was by **stagecoach**.
5. It cost the miners a lot of money to buy anything to cook in a **skillet**.
6. When silver was found, the small camp of Leadville became a **boom town**.
7. Mr. Grant said the miners found huge silver **nuggets**.

▶ **Imagine you live in the Old West. Write three sentences about your life using three vocabulary words.**

Possible response: I have just found some gold nuggets! The other miners can't believe it! I'm glad I settled here.

Books for All Learners Lessons, Grade 3 139

▲ Vocabulary Practice p. 139

Name ______________________

A Trip to the Past
Comprehension

▶ **Circle the best answer for each question.**

1. Which is the best definition of *greenhorns*?
   - A gold or silver shavings, nuggets, or bars
   - B miners who left the camps due to hunger
   - C another term for a boom town
   - (D) miners new to the diggings
2. Using pages 8-9, which of the following is the best example of self-questioning?
   - F How did a small town get so famous?
   - (G) Why did the miners stay in such conditions?
   - H Were the miners ever hungry?
   - J How often do most people visit the museum?
3. Which of the following is a fact found in the selection?
   - A Leadville must be an interesting place to visit.
   - B It must have been hard to live in Leadville.
   - (C) Mr. Grant tells the Kims about Leadville.
   - D Mr. Grant is a very friendly person.
4. Which of the following was NOT mentioned as a danger faced by miners?
   - F cholera or other diseases spread by rats
   - G dirty living conditions
   - H hunger
   - (J) poor weather conditions

▶ **Write an ad for Leadville using three facts from the story.**

Possible response: Come to Leadville! In only three years we have grown from 200 people to over 15,000! Why you ask? We have the largest source of silver in the world! People from all over come to Leadville, even from countries like Mexico and China! Don't miss out!

140 Books for All Learners Lessons, Grade 3

▲ Comprehension Practice p. 140

Name ______________________________

▶ **Write the word that best completes the sentence.**

miners	**landmark**	**settled**	**stagecoach**	**nuggets**
boom town	**skillet**			

1. ______________ left Leadville when the town ran out of silver.
2. People traveled for miles and ______________ in Leadville.
3. Mrs. Kim sees on a map that the General Store is the town's ______________.
4. One way miners traveled to Leadville was by ______________.
5. It cost the miners a lot of money to buy anything to cook in a ______________.
6. When silver was found, the small camp of Leadville became a ______________.
7. Mr. Grant said the miners found huge silver ______________.

▶ **Imagine you live in the Old West. Write three sentences about your life using three vocabulary words.**

____

____

____

____

Name ____

Circle the best answer for each question.

1. Which is the best definition of *greenhorns*?
A gold or silver shavings, nuggets, or bars
B miners who left the camps due to hunger
C another term for a boom town
D miners new to the diggings

2. Using pages 8-9, which of the following is the best example of self-questioning?
F How did a small town get so famous?
G Why did the miners stay in such conditions?
H Were the miners ever hungry?
J How often do most people visit the museum?

3. Which of the following is a fact found in the selection?
A Leadville must be an interesting place to visit.
B It must have been hard to live in Leadville.
C Mr. Grant tells the Kims about Leadville.
D Mr. Grant is a very friendly person.

4. Which of the following was NOT mentioned as a danger faced by miners?
F cholera or other diseases spread by rats
G dirty living conditions
H hunger
J poor weather conditions

Write an ad for Leadville using three facts from the story.

____

____

____

____

Name ______________________________

“They were hungry,” Mr. Grant said. “They ate salt pork, beef jerky, and 13
even spoiled bacon—anything to fill their growling stomachs.” 22

“Ugh!” Tracy and Tod said together. 28

“They were also dirty,” said Mr. Grant. “Dirty living conditions caused 39
terrible diseases. Rats scurried around at night spreading sickness. The 49
miners might try to hit the rats with a skillet or log, but many escaped. One 65
out of 25 miners in Sacramento died from a disease called cholera.” 77

“Didn’t the miners wash?” Tracy asked. 83

“Sunday was bath day,” Mr. Grant said. “They made their own soap by 96
boiling water, ashes, and bear fat in a big kettle.” 106

“Bear fat!” the kids yelled together. 112

“Bear fat was used for many things,” explained Mr. Grant. “Miners 123
rubbed bear fat into their leather boots to make them waterproof. Of course, 136
bear meat was a delicious treat for the hungry miners.” 146

“Food cost a lot in the boom towns of the 1870s,” Mr. Grant said. “A 161
pound of potatoes cost fourteen cents in New York, but one dollar in a boom 176
town.” 177

FLUENCY SCORE

Total Words Read Per Minute ________

Number of Errors − ________

Number of Words Read Correctly ________

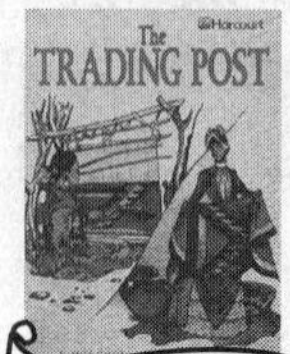

The Trading Post

by Meish Goldish

Historical Nonfiction

SUMMARY: Sara meets a Navajo girl named Maya at the trading post. They would like to be friends, but Maya can't speak English and Sara can't speak the Navajo language. This doesn't stop them from exchanging gifts and seeing one another.

- **Focus Skill:** Compare and Contrast
- **Focus Strategy:** Create Mental Images
- **Vocabulary:** trading, schooner, harvest, machete, pulp, bargain, support

Differentiated Instruction

Paired Reading Have students read aloud in tandem. If one reader makes a mistake, the partners must start at the beginning of the page. At the end of each chapter, have students ask their partners questions about the chapter. After the partners each answer two questions correctly, they may continue to the next chapter.

Prepare to Read

Preview/Set Purpose Have students preview the book by looking at the illustrations and the table of contents. Have students tell what they think the book will be about. Point out that this is a historical fiction book, which is a book about made-up people in a real place and time. Ask what they think they will learn from this book.

Compare and Contrast

Focus Skill

Teach/Model Display a compare-contrast chart similar to the one shown below. Tell students that good readers think about how things are alike and different to help them understand what they read. Ask students to give examples of items that have similarities and differences. Have students read pages 4 and 5 and try to find ways that Maya and Sara are alike and different to list on the chart.

Maya	Both	Sara
Navajo	Same age	European-American
confused about milk can	Visit the trading post	**understands Maya's confusion**

Practice/Apply Have partners add to their compare-contrast chart as they read *The Trading Post.*

Create Mental Images

Focus Strategy

Teach/Model Tell students that good readers create mental images to help them understand important ideas as they read. When good readers come across a description of an object or event in a story, they try to picture the object or event in their minds. Model this strategy for students by having them read page 2. Say, *When I read page 2, I imagine a store in the 1880s. I picture that the store is made from wood, with a tall roof and tall wooden shelves on all the walls. The shelves are stacked with shiny cans.* Point out that creating mental images can help them picture similarities and differences among people, places, and things in the selection.

Practice/Apply Have partners stop whenever they find a description and use the description to create a mental image. Have them share their mental image, and tell what details they used to create it.

Build Vocabulary

Help students use context to understand key vocabulary. Have them answer each question by reading the words and sentences that surround the word.

trading (p.2)	Does *trading* require paper money? **No, trading means exchanging one thing for another.**
schooner (p.3)	Is a *schooner* like a sailboat? **It has sails like a sailboat, but it is bigger.**
harvest (p.3)	What is another way to say that a farmer must *harvest* crops? **A farmer must gather crops.**
machete (p.3)	Is a *machete* like a kitchen knife? **No, it is heavy and used on plants.**
pulp (p.7)	Where would you find a fruit's *pulp*? **on the inside**
bargain (p. 3)	How might you *bargain* with a store owner? **Possible response: Offer him money.**
support (p. 14)	Can a tent *support* poles? **No, poles support the tent.**

CM 145 **Independent Practice:** Vocabulary Practice on page 145

Read and Comprehend

Monitor Comprehension

Guide students as they read the text, using the following questions:

Pages

2–3 Why did Mr. Hubbell let the farmer buy a machete with a model schooner? **Mr. Hubbell runs a trading post, where people trade items for other items.** INTERPRET INFORMATION

4–7 How did opening the can solve Maya's family's confusion? **They could see that the can held milk, and not flowers.** MAKE INFERENCES

8–11 Where does Sara get the idea for her gift?: **from the milk can** STORY EVENTS

Describe the red flowers that Sara gives Maya. **wild flowers that look like poppies** CREATE MENTAL IMAGES (Focus Strategy)

12–15 Why is Mr. Hubbell at the reservation? **to help people communicate** PROBLEM-SOLUTION

16 What differences do you notice between Maya and Sara's homes? **Sara lives in a house and Maya lives in a tent.** COMPARE AND CONTRAST (Focus Skill)

Think and Respond

Have students answer the questions on the inside back cover of the book.

1. **Sara is telling this story.**
2. **The milk cans were the same as the other cans because they had pictures on the front of them. They were different because the pictures were of flowers, not of the food inside the cans.**
3. **The story is mainly about how a girl named Sara makes friends with a Navajo girl named Maya.**
4. **Possible response: I learned that people in the Old West did not use money to get what they needed, but instead traded their goods.**
5. **The gifts are similar because both of the girls made them. They are different because one is a can of flowers, and the other is a blanket.**
6. **Possible response: I would like to shop at a trading post because it would be fun to trade items instead of use money to buy them.**

SOCIAL STUDIES

Social Studies Connection Have students draw maps of the area around their homes.

School-Home Connection Have students read *The Trading Post* with someone at home. Have them find out what Native American tribes are near their home.

CM 146 **Independent Practice:** Comprehension Practice on page 146

Write

Writer's Craft: Organization

Write a Research Report Remind students that good writers organize research reports into paragraphs. Each paragraph has a main idea sentence and supporting details. Have students turn to page 9 of *The Trading Post* and reread the second paragraph. Say, *The author is explaining how Sara's parents trade baskets and blankets for a saddle. The author organizes the details in time order.* Model how to fill in a main idea/details grid like the one below. Have students use a main idea/details grid and the text on page 14 to write a paragraph about Sara's trip to the Navajo reservation.

The Navajo Reservation		
Main Idea:		
Detail 1:	**Detail 2:**	**Detail 3:**

Fluency

Oral Reading Fluency

CM 147

Assessment Use Oral Reading Fluency Practice on page 147 to assess students' oral reading fluency.

Name ____________

The Trading Post
Vocabulary

▶ **Write the word that best matches the definition.**

harvest	pulp	machete	trading
schooner	support	bargain	

1. To gather a crop **harvest**.
2. The soft part of certain fruits and vegetables **pulp**.
3. To talk about a trade in order to get a better deal **bargain**.
4. To hold the weight of **support**.
5. A heavy knife used as a tool for cutting plants **machete**.
6. Used for the purpose of exchanging one thing for another **trading**.
7. A ship that has two or more masts with sails in the front and back **schooner**.

▶ **Read the paragraph below. Then fill in the vocabulary word that fits in the blank.**

The **trading** post has many interesting things to see. There are sacks of flour, barrels of pickles, and even a small toy **schooner** that really floats in water! After each fall's **harvest**, Father trades crops for food, cloth, and other things that we need. We don't have money for fancy things, but the sights and sounds of the trading post are a treat themselves!

▲ Vocabulary Practice p. 145

Name ____________

The Trading Post
Comprehension

▶ **Circle the best answer for the questions below.**

1. Why does Sara want to help Maya's family?
   - **A** She wants to visit the Navajo reservation and play with Maya.
   - **B** She works at the trading post and helps customers find things.
   - **C** She wants to stay at the trading post, not help her mother at home.
   - **(D)** She thinks trade is difficult for them since they don't speak English.
2. Which of these words describes both Maya and Sara?
   - **F** unhappy
   - **(G)** friendly
   - **H** talkative
   - **J** shy
3. Which of these sentences tells the main idea of the story?
   - **(A)** Real friendship can overcome many difficulties.
   - **B** Trading posts were the start of many pioneer friendships.
   - **C** Two blankets and two baskets are equal to one saddle.
   - **D** Maya's parents don't want her to see Sara again.

▶ **What parts of *The Trading Post* help you identify it as historical fiction? Why?**

Possible response: The story is fiction because Sara and Maya are made-up characters. The story is made up, too, but is in a real place and time. The date, 1883, helps me know that the story takes place long ago.

▲ Comprehension Practice p. 146

Name ______________________________

▶ **Write the word that best matches the definition.**

harvest	**pulp**	**machete**	**trading**
schooner	**support**	**bargain**	

1. To gather a crop ______________.
2. The soft part of certain fruits and vegetables ______________.
3. To talk about a trade in order to get a better deal ______________.
4. To hold the weight of ______________.
5. A heavy knife used as a tool for cutting plants ______________.
6. Used for the purpose of exchanging one thing for another ______________.
7. A ship that has two or more masts with sails in the front and back ______________.

▶ **Read the paragraph below. Then fill in the vocabulary word that fits in the blank.**

The ______________ post has many interesting things to see. There are sacks of flour, barrels of pickles, and even a small toy ______________ that really floats in water! After each fall's ______________, Father trades crops for food, cloth, and other things that we need. We don't have money for fancy things, but the sights and sounds of the trading post are a treat themselves!

Name ____

▶ **Circle the best answer for the questions below.**

1. Why does Sara want to help Maya's family?

A She wants to visit the Navajo reservation and play with Maya.

B She works at the trading post and helps customers find things.

C She wants to stay at the trading post, not help her mother at home.

D She thinks trade is difficult for them since they don't speak English.

2. Which of these words describes both Maya and Sara?

F unhappy

G friendly

H talkative

J shy

3. Which of these sentences tells the main idea of the story?

A Real friendship can overcome many difficulties.

B Trading posts were the start of many pioneer friendships.

C Two blankets and two baskets are equal to one saddle.

D Maya's parents don't want her to see Sara again.

▶ **What parts of *The Trading Post* help you identify it as historical fiction? Why?**

____

____

____

____

Name ______________________________

I'll never forget the day that one Navajo family came to the trading post. **14**
That was the day I met Maya. **21**

Maya and her parents were standing by a shelf of canned goods. They **34**
were staring at a can of milk for a very long time. I could tell they were **51**
puzzled. I walked over to help. **57**

"It's milk," I said. All three of them gave me a funny look and shook their **73**
heads. I realized they didn't understand me. I took the can and held it up to **89**
my mouth. **91**

"Milk," I repeated. "You drink it." **97**

Maya pointed to the picture on the cover of the can. It showed a bunch **112**
of pretty red flowers. Now I saw the problem! The family thought the can **126**
held flowers, not milk! **130**

I went to Mr. Hubbell and explained the problem. He opened the can. **143**
The family smiled when they saw the milk inside. **152**

Maya pointed to herself. "Maya," she said. **159**

I pointed to myself and said, "Sara." **166**

FLUENCY SCORE

Total Words Read Per Minute ____________

Number of Errors − ____________

Number of Words Read Correctly ____________

The Baker and the Rings

by Anne Miranda

Fairy Tale

SUMMARY When a giant steals a roomful of the queen's rings, a baker gets them back by baking the giant's favorite raisin bread. The baker gets the giant to agree that, for the next 20 days, he will pay double what he paid the day before for each loaf.

- **Focus Skill:** Main Idea and Details
- **Focus Strategy:** Adjust Reading Rate
- **Vocabulary:** congratulations, value, amount, receive, combination, choices

Differentiated Instruction

Paired Reading Organize students into pairs. Have them read the selection aloud to each other, alternating with each page. After students have finished reading, have them discuss in pairs a part of the story they found interesting.

Prepare to Read

Preview/Set Purpose Have students examine the cover and title. Ask them to preview the book by looking at the pictures. Ask them what the giant and the baker on the cover might be doing. Ask what they think they will learn from this book.

Focus Skill

Main Idea and Details

Teach/Model Display a chart like the one below. Explain that good readers organize a text into a main idea and supporting details to help them understand what they read. Tell students that the main idea is the most important idea in a story or part of a story. The details give more information to support the main idea. Read page 2 with students. Say, *The main idea in this passage is that all the queen's gold rings were gone. A detail is that the butcher, baker, and candlestick maker heard her cry. This gives more information about the main idea.*

Main Idea: The queen's gold rings were gone.
Detail: The butcher, baker and candlestick maker heard her cry.

Practice/Apply Have partners identify other supporting details on page 2 and add them to the chart above.

Focus Strategy

Adjust Reading Rate

Teach/Model Explain that good readers slow down when they come to a difficult part of the text. Tell students that adjusting their reading rate in this way can help them understand difficult or confusing passages. Have students read page 8. Say, *The deal the baker makes with the giant seems complicated. I'm going to read this passage more slowly to make sure I understand.* Point out that adjusting the reading rate to remember details can also help readers gain a better understanding of important ideas.

Practice/Apply Have partners slow down when they come to difficult parts of the text.

Build Vocabulary

Help students use context to understand key vocabulary. Have them answer each question by reading the words and sentences that surround the word.

congratulations (p. 16)	When was the last time someone told you *congratulations*? **Possible response: when I did well on a test**
value (p. 7)	What is something of *value* to you? **Possible response: my baseball cards**
amount (p. 13)	What *amount* of money might you pay for an ice cream cone? **a few dollars**
receive (p. 16)	When might someone *receive* a card? **his or her birthday**
combination (p. [illegible])	What *combination* of food tastes good? **Possible response: chocolate and peanut butter; cheese and tomatoes**
choices (p. 7)	What *choices* have you made today? **Possible response: what to eat or wear**

CM 151 **Independent Practice:** Vocabulary Practice page 151

Read and Comprehend

Monitor Comprehension

Guide students as they read the text, using the following questions:

Pages

2–3 How do people know the queen is upset? **They hear her cry.** MAKE INFERENCES

4–5 What do you think the giant will do? **Possible response: trade the rings for bread** MAKE PREDICTIONS

6–9 What is the main idea of page 7? **People must make choices.** MAIN IDEA AND DETAILS (Focus Skill)
Why does the baker ask the giant to make a promise? **to make sure he gives the rings back** CHARACTER'S MOTIVATIONS

10–12 Would you keep buying the bread? Why or why not? **Possible response: No; it costs too much.** IDENTIFY WITH CHARACTERS
How many rings did the giant pay on Day 19? **262,144 rings** ADJUST READING RATE (Focus Strategy)

Think and Respond

Have students answer the questions on the inside back cover of the book.

1. **This is not a true story because it is a fairy tale.**
2. **These details include the special ingredients in the bread, the difficult steps the baker takes to prepare the bread, the description of the smell of the bread, and the way the giant is willing to do anything for the bread.**
3. **The main idea of the story is that keeping promises is important.**
4. **Possible response: I learned that making bargains can work if each person wants what the other has.**
5. **The giant's attitude towards the baker is the same at the beginning and end of the story because he likes the bread that she makes and wants to eat it. It is different because at the beginning of the story he thinks he is getting a better deal than the baker, but later he is mad at her for tricking him.**
6. **Possible response: I do think it is important to keep a promise so that people can trust each other.**

Math Connection Have students figure out the cost of a baseball card if it has doubled its price from $.60. Then have them figure out the new cost of 2, 3, and 4 baseball cards.

MATH

School-Home Connection Ask students to have a family member help them draw something they would like to have. Have them draw something they would trade for it.

CM 152 **Independent Practice:** Comprehension Practice page 152

Write

Writer's Craft: Organization

Write a Research Report Tell students that good writers organize their ideas. Say, *Look at page 5. The ideas in this paragraph are clearly organized. The main idea comes first, and then the supporting details are listed in time order.* The best way to organize ideas for a research report is with an outline. Explain to students that an outline is a map for a report. Writers come up with ideas and topics for a report. Then, they begin their paragraphs with a topic sentence. They finish the outline with supporting details. Have students research the history of raisins. Have them use the organizer below to write an outline.

Title: The History of Raisins

1. Introduction: ____________
2. Middle: ____________
3. Conclusion: ____________

Fluency

Oral Reading Fluency

CM 153

Assessment Use Oral Reading Fluency Practice page 153 to assess students' oral reading fluency.

Name ____________

The Baker and the Rings
Vocabulary

▶ **Write the word that best completes each sentence.**

congratulations receive amount choices combination value

Dear Jamal,

You have won the school writing contest, and I am happy to say **congratulations** to you! I know you did a great **amount** of work on your report. You might have wanted to spend your time playing, but you made good **choices** and did your work instead. I really liked the **combination** of pictures and charts that you put together. You will **receive** a prize from the school. I hope you will think it has **value**.

Sincerely,
Principal Tan

▶ **Write a thank you letter to someone who has helped you. Use at least three vocabulary words.**

Possible response: Dear Mom, Thank you for helping me make good choices in my life. You have more value to me than any amount of baseball cards. I will never forget your advice.

Vocabulary Practice p. 151

Name ____________

The Baker and the Rings
Comprehension

▶ **Underline the details that support the main idea, "The baker baked very well."**

A. The queen was upset that her rings were gone.
B. The giant came when he smelled the bread.
C. The giant left footprints outside the castle window.
D. The giant paid whatever the baker asked.
E. The giant did not get tired of the bread.

▶ **Answer the questions below.**

1. **What thing of value did the queen give to get her rings back? To whom did she give it?**
   She gave the baker the chance to be a princess.
2. **How did the baker outsmart the giant?**
   She got the giant to agree to pay double each day for her bread. The giant did not realize how much the total would be.
3. **How many rings did the giant pay on Day 12? How do you know?**
   He paid 2,048 rings. According to the baker's chart on page 12, he paid 512 rings on Day 10. That means he paid 1,024 on Day 11, and 2,048 on Day 12.

Comprehension Practice p. 152

Name ____

▶ Write the word that best completes each sentence.

congratulations receive amount choices combination value

Dear Jamal,

You have won the school writing contest, and I am happy to say

__________________ to you! I know you did a great

__________________ of work on your report. You might have wanted to spend your time playing, but you made good

__________________ and did your work

instead. I really liked the __________________ of pictures and

charts that you put together. You will __________________ a prize

from the school. I hope you will think it has __________________.

Sincerely,

Principal Tan

▶ Write a thank you letter to someone who has helped you. Use at least three vocabulary words.

____

____

____

____

____

____

Name ____________________

▶ **Underline the details that support the main idea, "The baker baked very well."**

A. The queen was upset that her rings were gone.
B. The giant came when he smelled the bread.
C. The giant left footprints outside the castle window.
D. The giant paid whatever the baker asked.
E. The giant did not get tired of the bread.

▶ **Answer the questions below.**

1. What thing of value did the queen give to get her rings back? To whom did she give it?

____

____

2. How did the baker outsmart the giant?

____

____

3. How many rings did the giant pay on Day 12? How do you know?

____

____

____

____

Name ______________________________

Each day the giant came to the kitchen, and each day there was a 14
delicious hot loaf of bread waiting for him. Each day the giant had to pay 29
the baker double the number of rings he had paid the day before. 42

The baker kept a chart of the number of rings the giant had to pay each 58
day and the total number he had paid. 66

In two weeks, the number of rings the giant had to pay had reached a 81
rather hefty amount. Because the giant loved the baker's bread so much, he 94
always paid whatever the baker asked. On Day 14, he paid the baker 8,192 108
rings. The next day, the amount doubled to 16,384. On Day 16, the giant 122
paid the baker 32,768 rings. On Day 17, the number of rings he had to pay 138
was 65,536! 140

The giant needed a wheelbarrow to carry all the rings to the castle. The 154
baker needed help to count the rings. The butcher and the candlestick 166
maker were happy to help. 171

FLUENCY SCORE

Total Words Read Per Minute ______________

Number of Errors − ______________

Number of Words Read Correctly ______________

City Celebrations

by Debra Hershkowitz

SUMMARY This story, written in the form of a young boy's travel journal, describes visits to six big cities throughout the United States and highlights their special celebrations and traditions. It ends by describing an event celebrated by Americans everywhere.

- **Focus Skill:** Realistic fiction
- **Focus Strategy:** Create Mental Images
- **Vocabulary:** admiring, choosy, average, signal, tracks, celebrations

Differentiated Instruction

Paired Reading Organize students in pairs. Have them read the selection aloud to each other, alternating with each page. After students have finished reading, have them discuss in pairs a part of the story they found interesting.

Prepare to Read

Preview/Set Purpose Have students preview the book by examining the cover, title, and pictures. Point out that this selection is a collection of journal entries, in which the cities mentioned are actual places in our country. Have students use the information they gathered from previewing the pictures to predict what they will learn about each city in the story.

Summarize

Focus Skill

Teach/Model Remind students that to summarize means to retell the important events and ideas from a selection in their own words. Discuss with students what makes a good summary. Explain that the summary should be shorter than the original selection. It should include only the most important events or ideas, and it should not include the opinions of the person who is summarizing. Tell students to read the first paragraph on page 2. Say, *To summarize this paragraph, I would say: The New York City Marathon is a 26 mile race. Jamie's mother runs it every year.*

Practice/Apply After students have read *City Celebrations*, have them choose one journal entry and summarize it.

Create Mental Images

Focus Strategy

Teach/Model Tell students that good readers create mental images to help them imagine events as they read a story. Have students read pages 2 and 3. Say, *When I read the first entry, I notice that it refers to New York City. Both pictures show the marathon, but no pictures show famous spots in the city like the Statue of Liberty. I decide to form pictures in my mind about this landmark since I have never been to New York City.*

Practice/Apply Ask students to stop when they read an interesting part of the story that is not pictured and have them use the Creating Mental Images strategy. Invite students to share their mental images with the class.

Build Vocabulary

Help students use context to understand key vocabulary. Have them answer each question by reading the words and sentences that surround the word.

admiring (p. 14)	Why are people seen *admiring* diamonds? **Diamonds are beautiful.**
choosy (p. 9)	Why are *choosy* eaters hard to please? **They like only to eat certain things.**
average (p. 2)	Why might an *average* day be boring? **Nothing special happens.**
signal (p. 11)	How do you *signal* you want the teacher to call on you? **I raise my hand.**
tracks (p. 5)	Which animal might leave tiny *tracks* in the dirt? **Possible response: squirrel**
celebrations (p. 4)	What do *celebrations* honor? **holidays, achievements, or happy occasions**

CM 157 **Independent Practice:** Vocabulary Practice page 157

Read and Comprehend

Monitor Comprehension

Guide students as they read the text, using the following questions

Pages

2–5 Write a sentence that summarizes Jamie's first experience of the New York Marathon. **Possible response: Jamie enjoys watching all the interesting people who run in the marathon.** SUMMARIZE (Focus Skill)

What language is Jamie speaking at the end of his entry when he says, "Gung hay fat choy!"? **Chinese** MAKE INFERENCES

6–9 What monument is pictured on page 9? **The Washington Monument** USE GRAPHIC AIDS

10–13 Do you think Jamie liked the cow statues? How can you tell? **Yes, he smiled as they got closer.** MAKE JUDGMENTS

14–15 Do you think the man will make any money playing his guitar in the subway station? **Possible response: Yes, people like music.** PREDICT OUTCOMES

16 What form of writing does Jamie use? **journal writing** NARRATIVE ELEMENTS

Think and Respond

Have students answer the questions on the inside back cover of the book.

1. **The writer uses the form of a diary.**
2. **Possible response: Watching the New York City Marathon for the first time was exciting.**
3. **Possible response: The main idea of this book is that all cities are unique and fun to visit.**
4. **Possible response: I learned that big cities have interesting things to see and do.**
5. **Possible response: Washington, D.C., is like San Francisco because both cities have special celebrations: Chinese New Year in San Francisco and the Cherry Blossom Festival in Washington, D.C. The cities are different because Washington, D.C., has the White House and other United States government buildings.**
6. **Possible response: I would like to visit Washington, D.C., so I could see the museums.**

SOCIAL STUDIES

Social Studies Connection Have students find more information about one of the cities mentioned in the selection.

School-Home Connection Tell students to ask a family member to describe his or her favorite town or city.

CM 158 **Independent Practice:** Comprehension Practice page 158

Write

Writer's Craft: Word Choice

Write a Poem Tell students that good writers create a picture in the reader's mind when they write poetry. Point out that poets use colorful, vivid language to show readers how things look, sound, taste, feel, and smell. Have students write an acrostic poem about their hometown. Remind students that acrostic poems spells the word (city name) down the left side of the paper and that each line of the poem must start with the letter on that line. Encourage them to address as many of the senses in their poem as they can.

My Acrostic

Fluency

Oral Reading Fluency

CM 159 **Assessment** Use Oral Reading Fluency Practice page 159 to assess students' oral reading fluency.

Name ______________________

City Celebrations
Vocabulary

▶ **Match each word to its definition.**

C 1. admiring
D 2. choosy
E 3. average
F 4. signal
A 5. tracks
B 6. celebrations

A. prints left in the ground
B. joyous events that honor holidays
C. looking at with wonder
D. picky, hard to please
E. ordinary, not special
F. to make known, to announce

▶ **Draw a picture of a mental image you created of one of the cities while reading the story. Underneath your illustration write a description that uses at least two of the vocabulary words from above.**

Possible response: The people are admiring the flowers in Washington, D.C. Today is not an average day.

Books for All Learners Lessons, Grade 3 157

▲ Vocabulary Practice p. 157

Name ______________________

City Celebrations
Comprehension

▶ **Circle the best answer to the questions below.**

1. How many cities does Jamie mention during the story?
   A 4
   B 7
   C 8
   (D) 6
2. Which city that Jamie visited is the capital of our country?
   (F) Washington, D.C.
   G St. Louis
   H New York
   J Boston
3. What celebration does Jamie observe while in California?
   A Independence Day
   B Mom's birthday
   (C) Chinese New Year
   D National Cherry Bloom Festival
4. Which of the following is NOT stated in Jamie's entries?
   (F) Chicago is Jamie's favorite city they visited.
   G It is best to visit Washington, D.C. in the spring.
   H The Chinese New Year is not celebrated in January.
   J Jamie's mom runs in the New York City Marathon each year.

▶ **Respond to the prompt below using complete sentences.**

5. Summarize Jamie's visit to Boston in *City Celebrations.*
   Possible response: Jamie visited his aunt and uncle. He took a ride on a boat and saw a woman walking eight dogs.

158 Books for All Learners Lessons, Grade 3

▲ Comprehension Practice p. 158

Name ______________________________

Match each word to its definition.

____ **1.** admiring	**A.** prints left in the ground
____ **2.** choosy	**B.** joyous events that honor holidays
____ **3.** average	**C.** looking at with wonder
____ **4.** signal	**D.** picky, hard to please
____ **5.** tracks	**E.** ordinary, not special
____ **6.** celebrations	**F.** to make known, to announce

Draw a picture of a mental image you created of one of the cities while reading the story. Underneath your illustration write a description that uses at least two of the vocabulary words from above.

____

____

____

____

Name ______________________________

▶ **Circle the best answer to the questions below.**

1. How many cities does Jamie mention during the story?

A 4
B 7
C 8
D 6

2. Which city that Jamie visited is the capital of our country?

F Washington, D.C.
G St. Louis
H New York
J Boston

3. What celebration does Jamie observe while in California?

A Independence Day
B Mom's birthday
C Chinese New Year
D National Cherry Bloom Festival

4. Which of the following is NOT stated in Jamie's entries?

F Chicago is Jamie's favorite city they visited.
G It is best to visit Washington, D.C. in the spring.
H The Chinese New Year is not celebrated in January.
J Jamie's mom runs in the New York City Marathon each year.

▶ **Respond to the prompt below using complete sentences.**

5. Summarize Jamie's visit to Boston in *City Celebrations.*

____

____

____

Name ______________________________

Dear Diary, 2

My city has a big race called the New York City Marathon. It's about 16
26 miles long. My mother runs it every year. She says she's not very fast 31
and that she is an average runner. But I wouldn't call anyone who runs 45
26 miles "average"! That's even a long way to walk. 55

I remember the first time we saw Mom run. It was so exciting! I sat 70
on Dad's shoulders, and he kept saying, "Look! Look!" I got dizzy from 83
watching all the runners go by. My hands got tired from clapping. 95

Some runners were wearing weird costumes. One person was dressed 105
as a rhinoceros. He wore a sign that said *Save the rhinos.* There was 119
another person dressed as a waiter. He was carrying food on a tray. 132

There were blind runners who had partners to guide them. Also, 143
a lot of racers weren't running. Some racers were in wheelchairs, and 155
some were on crutches. I was so excited! 163

FLUENCY SCORE

Total Words Read Per Minute ____________

Number of Errors — ____________

Number of Words Read Correctly ____________

A Circle Story

by Meish Goldish

SUMMARY In the 1940s a girl named Lisa loans her favorite book to a classmate named Miguel. The book is filled with notes Lisa has written about the book. When Miguel's father is suddenly transferred to another Army base, Lisa thinks she will never see her cherished book again. Sixty years later, Lisa is delighted when she receives her book in the mail. It is stuffed with notes and reflections written by three generations of Miguel's family.

- **Focus Skill:** Cause and Effect
- **Focus Strategy:** Read Ahead
- **Vocabulary:** windmill, cherished, furrows, ample, shunned, growth

Differentiated Instruction

Paired Reading Organize students in pairs. Have them read the selection aloud to each other, alternating with each page. After students have finished reading, have them discuss in pairs a part of the story they found interesting.

Prepare to Read

Preview/Set Purpose Have students preview the book by reading the chapter headings and looking at the pictures. Have them read the title. Explain that this is a fiction story, in which the plot, characters, and setting are not real. Have students predict what the story is about.

Cause and Effect

Focus Skill

Teach/Model Display a chart similar to the one below. Explain that a cause is an event that makes something happen, and an effect is what happens. Authors often use signal words, such as *because* or *as a result of*, to show a cause-effect relationship. Have students read this sentence: *She wasn't ready to sleep yet, so she sat up in bed, reading her book.* Say, She wasn't ready to sleep yet *is the cause because it makes something happen*; she sat up in bed, reading her book *is the effect because this is what happens.*

Cause	Effect
She wasn't ready to sleep yet.	She sat up in bed, reading her book.

Practice/Apply After they have finished reading tell partners to add one cause and effect relationship to the chart.

Read Ahead

Focus Strategy

Teach/Model Tell students that good readers use the Read Ahead strategy to help them understand a selection. Explain that reading ahead can help clear up confusion. Have students read pages 2 – 3. Say, *I don't understand what the word* champion *means. I read ahead and learn that Lucky could jump higher and run faster than the other horses. So* champion *must mean* someone who is the best.

Practice/Apply Have partners use the Read Ahead strategy to find the meaning of *transferred* on page 11.

Build Vocabulary

Help students use context to understand key vocabulary. Have them answer each question by reading the words and sentences that surround the word.

windmill (p. 6)	What might a *windmill* be used for? **Possible response: to make power**
cherished (p. 2)	What have you *cherished*? **Possible response: dog**
furrows (p. 5)	Who makes *furrows* in the soil? **Possible response: farmer**
ample (not in text)	If you had *ample* food, what does it mean? **I had more than enough.**
shunned (p. 3)	If you *shunned* someone, what did you do? **stayed away from them**
growth (p. 15)	What *growth* might you find in a desert? **Possible response: shrubs, cacti**

CM 163 **Independent Practice:** Vocabulary Practice page 163

Read and Comprehend

Monitor Comprehension

Guide students as they read the text, using the following questions:

Pages

2–3 What did the farmer do when Lucky was too old to work? **He took Lucky to the school so children could ride him.** SEQUENCE

4–6 On what date did Lisa write her last note? **6/12/42** USE GRAPHIC AIDS

7–9 How does reading ahead help you define the word tucked? **Possible response: puts notes in the book each time, picture also shows what *tucking* looks like** READ AHEAD (Focus Strategy)

10–11 Why was Lisa surprised when she went to Miguel's house? **There was a "vacant" sign outside his house.** CAUSE-EFFECT (Focus Skill)

12–16 What subject do you think Lisa studied at college? **Possible response: English Literature, because she liked to read.** DRAW CONCLUSIONS

How do you think Lisa felt when she got her book back? **She was happy.** CHARACTER'S EMOTIONS

Think and Respond

Have students answer the questions on the inside back cover of the book.

1. **Possible response: The chapter titles helped me figure out what was going to happen in the story.**
2. **She wasn't able to read the story because the boy who had borrowed the book had never returned it.**
3. **It is called *A Circle Story* because the book *A Horse Called Lucky* travels in a circle from Lisa to the Gonzales family and back to Lisa again.**
4. **I learned that sharing things with friends can make you and your friends happy.**
5. **Lucky and the farmer are alike because they both like to walk slowly through the fields and they both enjoy the outdoors. They are different because Lucky is a horse and the farmer is a man.**
6. **Possible response: I have read *Ramona Forever* and would read it again because I like funny books.**

Language Arts Connection Have students reread their favorite book and write notes on slips of paper. Have them share their notes with the class and say why they like the book.

School-Home Connection Have children ask a family member if they have ever given a favorite book to someone and not had it returned.

CM 164 **Independent Practice:** Comprehension Practice page 164

Write

Writer's Craft: Word Choice

Write a Thank-you Letter Say, *The words* surprised, tucked, *and* joked *on page 8 are examples of vivid words because they describe the character's thoughts.* Tell students that good writers use colorful, vivid language to show readers exactly what they mean. Tell students to pretend they are Lisa. Have them write a letter to Maria, Miguel's granddaughter, thanking her for returning the book. Explain that a thank-you letter is one kind of friendly letter. It thanks a person for giving a gift or doing something nice.

Address and Date:

Dear Ms.:

Body:

Conclusion:

Sincerely,:

Signature:

Fluency

Oral Reading Fluency

CM 165 **Assessment** Use Oral Reading Fluency Practice page 165 to assess students' oral reading fluency.

Name ______________________

A Circle Story
Vocabulary

▶ **Circle the letter of the word that describes or relates to the word in dark print.**

1. **growth**
   Ⓐ plants **B** chairs
2. **furrows**
   Ⓐ soil **B** water
3. **ample**
   A little Ⓑ lots
4. **shunned**
   Ⓐ ignore **B** greet
5. **windmill**
   Ⓐ power **B** sunlight
6. **cherished**
   A hated Ⓑ loved

▶ **Write a short paragraph about your favorite book using three vocabulary words.**

Possible response: My most cherished book is *Where the Red Fern Grows*. It is about the growth of a boy and his two dogs. Ample people love this book because anyone can relate to it. The boy and his dogs are the perfect team. Before the dogs came to live with him, they were shunned because one was the runt of the litter.

© Harcourt

Books for All Learners Lessons, Grade 3 163

▲ Vocabulary Practice p. 163

Name ______________________

A Circle Story
Comprehension

▶ **Reread the story. Answer the questions below.**

1. Name the two main characters in this story.
   Lisa and Miguel
2. Why is the second chapter called *The Loan*?
   Lisa lends her book to Miguel.
3. What color is Lucky's coat? How do you know?
   Brown and white; the note Lisa wrote on page 5 says Lucky's coat reminds her of vanilla-fudge ice cream.
4. Read page 16 and write the names of Miguel's grandchildren.
   Maria, Mike, Amy, Jason, and Elvira
5. What is the purpose of a family tree?
   to show the different generations of a family

▶ **Complete the chart with the correct cause or effect.**

Cause	Effect
Lucky got old and couldn't do work.	The farmer's son buys a tractor.
Students in Lisa's class were asked to talk about their favorite book.	**Lisa gave a report on *A Horse Called Lucky*.**
She did not want to forget the thoughts that crossed her mind when she read the book.	Lisa wrote notes on small slips of paper.
Miguel's father was transferred to another army base.	**Miguel left school quickly and could not return Lisa's book.**

© Harcourt

164 Books for All Learners Lessons, Grade 3

▲ Comprehension Practice p. 164

Name ____

▶ **Circle the letter of the word that describes or relates to the word in dark print.**

1. **growth**
   A plants B chairs

2. **furrows**
   A soil B water

3. **ample**
   A little B lots

4. **shunned**
   A ignore B greet

5. **windmill**
   A power B sunlight

6. **cherished**
   A hated B loved

▶ **Write a short paragraph about your favorite book using three vocabulary words.**

____

____

____

____

____

____

Name ______________________________

Reread the story. Answer the questions below.

1. Name the two main characters in this story.

 ____

2. Why is the second chapter called *The Loan*?

 ____

3. What color is Lucky's coat? How do you know?

 ____

 ____

4. Read page 16 and write the names of Miguel's grandchildren.

 ____

5. What is the purpose of a family tree?

 ____

Complete the chart with the correct cause or effect.

Cause	Effect
	The farmer's son buys a tractor.
Students in Lisa's class were asked to talk about their favorite book.	
	Lisa wrote notes on small slips of paper.
Miguel's father was transferred to another army base.	

Name ____________________

Lisa heard the army base bugler play taps and knew that it was her 14
bedtime. She wasn't really ready to sleep yet, so she sat up in the bed, 29
reading her book. Lisa cherished this book more than any other. The title 42
was *A Horse Called Lucky.* 47

The story was about a horse that had been a racing champion. He jumped 61
higher and ran faster than other horses. 68

Then a farmer bought Lucky and the two plowed the cornfields together. 80
As they got older, they couldn't work as much, so the farmer's son bought a 95
tractor. 96

After that, the farmer and Lucky spent many days together, walking 107
slowly through the fields, enjoying the outdoors. Then, the farmer had an 119
idea. Just past the farm field was a school full of children. On Fridays, he 134
could take Lucky to the school and give the children rides. 145

While other work horses had shunned the school children, Lucky loved 156
all the attention. He was happy. 162

FLUENCY SCORE

Total Words Read Per Minute ____________

Number of Errors – ____________

Number of Words Read Correctly ____________

The Perfect Ending

by Vita Jimenez

SUMMARY The Davis family takes a weekend camping trip to a nearby national park, where they sleep under the stars and enjoy a long hike in the woods. On their way back to the campsite, a powerful earthquake startles them and opens up an underground cave, which they report to a park ranger. Weeks later, the family receives photos of the inside of the cave they discovered.

- **Focus Skill:** Locate Information
- **Focus Strategy:** Adjust Reading Rate
- **Vocabulary:** edges, range, peak, epicenter, magma, coast

Differentiated Instruction

Paired Reading Organize students in pairs. Have them read the selection aloud to each other, alternating with each page. After students have finished reading, have them discuss in pairs a part of the story they found interesting.

Prepare to Read

Preview/Set Purpose Have students preview the book by examining the cover and title. Point out that this selection is realistic fiction, in which the characters, setting, and plot could all be real. Have students set a purpose for reading, such as *I want to find out what the perfect ending is.*

Locate Information

Focus Skill

Teach/Model Say, *you can find information in many different parts of a book, including the title page, illustration, and captions.* Draw on the board a story map like the one below. Have students read page 2 and work with them to fill in the story map.

Characters	Settings	Events
Davis family	national park	**earthquake**
the Park Ranger	**the family's house**	**receiving photos**

Practice/Apply Have students fill in their story maps after they have finished reading.

Adjust Reading Rate

Focus Strategy

Teach/Model Tell students that good readers often change the rate, or speed, of their reading to match the text they are reading. Read pages 2–3. Say, *When I read pages 2–3, at first I am overwhelmed by all of the details the author is giving about the family and their trip. However, when I go back to the beginning and reread the text slower, I understand more. Reading complicated passages more slowly will help you understand new words and ideas.*

Practice/Apply Remind students to adjust the rate, or speed, of their reading to match the text as they read the story.

Build Vocabulary

Help students use context to understand key vocabulary. Have them answer each question by reading the words and sentences that surround the word.

edges (p. 7)	Which shape has no *edges*: a circle, a square, or a triangle? **a circle**
range (p. 4)	What is the closest mountain *range* to your school? **Possible response: the Rockies**
peaks (p. 4)	What word or words mean the same as *peaks*? **crests, tops, climaxes**
epicenter (not in text)	What is the *epicenter* of our school? **Possible responses: the cafeteria, the principal's office**
magma (p. 7)	Is *magma* hot or cold when it comes out of the volcano? **very hot**
coast (p. 3)	What ocean borders the California *coast*? **The Pacific Ocean**

CM 169 **Independent Practice:** Vocabulary Practice page 169

Read and Comprehend

Monitor Comprehension

Guide students as they read the text, using the following questions:

Pages

2–5 Do you think the family is looking forward to the camping trip? **Yes, they all love the outdoors.** FACT AND OPINION

6–7 If the rocks the man found were formed by magma, where do you think he may have found them? **around a volcano** MAKE INFERENCES

8–9 Why do you think the raven was making the noise? **Possible response: It could sense an earthquake was coming** MAKE INFERENCES

10–11 What important safety precautions did the author want you to learn? **During an earthquake, move away from trees and get to an open area.** AUTHOR'S PURPOSE

12–16 How could a cave have appeared that quickly? **Possible response: Many changes can happen to the land when the earth's crust moves and shakes.** CAUSE AND EFFECT
What were in the pictures that the ranger sent the Davis family? **rock formations** LOCATE INFORMATION (Focus Skill)

Think and Respond

Have students answer the questions on the inside back cover of the book.

1. **This story takes place at a campsite in a national park.**
2. **You learn on page 10 that the earthquake struck, in late afternoon.**
3. **Trees toppled over, and a cave opened up.**
4. **Possible response: I learned to bring a camera, binoculars, backpack, notebook, and pen.**
5. **Possible response: This story is like another one I read because the family learns about new things on their trip. It is different because they experience an earthquake on their trip.**
6. **Possible response: I would have gone to the clearing with them to get away from the trees.**

SCIENCE

Science Connection Have students make a list of facts they know about caves.

School-Home Connection Have students discuss with a family member what to do in case of an earthquake or other emergency.

CM 170 **Independent Practice:** Comprehension Practice page 170

Write

Writer's Craft: Word Choice

Write a Play Tell students that good writers use colorful, vivid language to show readers exactly what they mean. Explain that using active verbs, specific nouns, and descriptive adjectives are a great way to keep the reader's interest. Remind students that plays have dialogue between characters and stage directions that describe the time and place. Write a play about a family going on vacation. Use the family members from *The Perfect Ending* as the main characters.

Family Vacation

Fluency

Oral Reading Fluency

CM 171

Assessment Use Oral Reading Fluency Practice page 171 to assess students' oral reading fluency.

Name ______________________

The Perfect Ending
Vocabulary

▶ **Fill in each blank with the correct word.**

edges | range | peak | epicenter | magma | coast

1. There is a wide **range** of people with different heights in my class.
2. When a volcano erupts, hot **magma** comes out the top and the vents in the sides.
3. The tallest mountain in the world is Mt. Kilimanjaro in Tibet, whose **peak** is 19,340 feet in the air!
4. To save space, I like to write all the way to the **edges** of my paper.
5. The city of Los Angeles is on the **coast** of Southern California.
6. San Francisco was at the **epicenter** of two earthquakes: one in 1906, and one in 1989.

▶ **Write a postcard from one of the family members in *The Perfect Ending*. Use at least two of the vocabulary words from above.**

Possible response: Dear Grandma, We are having so much fun on vacation! We can see a mountain peak from our campsite. We met a man today who had a rock made out of magma. It was so cool.

▲ Vocabulary Practice p. 169

Name ______________________

The Perfect Ending
Comprehension

▶ **Reread the story. Circle the best answer to the questions below.**

1. Where did the Davis' go for their weekend trip?
   - A a historic site
   - B a beach resort
   - C a quiet place that has mountains
   - (D) a nearby national park
2. What things did the children bring along for the car ride?
   - F CDs and DVDs.
   - G word searches and games
   - (H) books and puzzles
   - J cell phones
3. What happened first?
   - (A) They met a fellow camper.
   - B They felt the earthquake.
   - C They talked to the park ranger.
   - D They discovered a cave.
4. What tool does Timmy use to see objects that are far away?
   - (F) binoculars
   - G thermometer
   - H seismograph
   - J computer

▶ **Where did you locate information about ravens?**

Possible response: I found information about ravens on page 9.

▲ Comprehension Practice p. 170

Name ______________________________

▶ **Fill in each blank with the correct word.**

edges	**range**	**peak**	**epicenter**	**magma**	**coast**

1. There is a wide ______________ of people with different heights in my class.

2. When a volcano erupts, hot ______________ comes out the top and the vents in the sides.

3. The tallest mountain in the world is Mt. Kilimanjaro in Tibet, whose ______________ is 19,340 feet in the air!

4. To save space, I like to write all the way to the ______________ of my paper.

5. The city of Los Angeles is on the ______________ of Southern California.

6. San Francisco was at the ______________ of two earthquakes: one in 1906, and one in 1989.

▶ **Write a postcard from one of the family members in *The Perfect Ending*. Use at least two of the vocabulary words from above.**

____

____

____

____

____

Name ______________________________

Reread the story. Circle the best answer to the questions below.

1. Where did the Davis' go for their weekend trip?
   - **A** a historic site
   - **B** a beach resort
   - **C** a quiet place that has mountains
   - **D** a nearby national park

2. What things did the children bring along for the car ride?
   - **F** CDs and DVDs
   - **G** word searches and games
   - **H** books and puzzles
   - **J** cell phones

3. What happened first?
   - **A** They met a fellow camper.
   - **B** They felt the earthquake.
   - **C** They talked to the park ranger.
   - **D** They discovered a cave.

4. What tool does Timmy use to see objects that are far away?
   - **F** binoculars
   - **G** thermometer
   - **H** seismograph
   - **J** computer

Where did you locate information about ravens?

____

____

____

Name ______________________________

The Davis family was getting ready for a weekend trip. They were 12
loading their things into the car, and although everyone had promised not to 25
pack too many things, the car was bulging with all their gear. 37

Mr. and Mrs. Davis were taking their two children, Timmy and Kerri, 49
camping for a weekend in a national park. The whole family loved to 62
explore the outdoors, and they were excited about their trip. 72

"Well, that about does it," said Dad, as he squeezed the last suitcase into 86
the trunk. "Let's go!" 90

Kerri and Timmy had brought things to keep them busy along the way. 103
Kerri read a book about horses, while Timmy worked on a tricky crossword 116
puzzle and asked for help every few minutes. 124

Mom and Dad took turns driving along the coast to the campground. 136

It was sunset when the family arrived at the campsite. In the distance, 149
they could see a beautiful mountain range. Pine trees covered all of 161
the peaks. 163

FLUENCY SCORE

Total Words Read Per Minute ____________

Number of Errors − ____________

Number of Words Read Correctly ____________

Amazing Armadillos

by Lisa Trumbauer

SUMMARY This informational book about armadillos begins with a legend about how armadillos came to be. It goes on to explain how long armadillos have been on the earth, how their unusual bodies are constructed, how they build homes and protect themselves, and where these small armored survivors live today.

- **Focus Skill:** Cause and Effect
- **Focus Strategy:** Reread to Clarify
- **Vocabulary:** eventually, converse, continent, sphere, universe, homeward

Differentiated Instruction

Paired Reading Organize students in pairs. Have them read the selection aloud, alternating with each page. After students have finished reading, have them discuss in pairs a part of the story they found interesting.

Prepare to Read

Preview/Set Purpose Have students preview the book by examining the cover and title. Point out that *Amazing Armadillos* is a nonfiction book, meaning it contains facts and information. Have students preview the illustrations. Ask them why they think pages 2–6 have drawings of people, while the rest of the pictures are photos and drawings of armadillos. Have students predict what the book will be about.

Cause and Effect

Focus Skill

Teach/Model Display a chart like the one below. Explain to students that a cause is the reason something happens, and an effect is the result of an action or event. Remind students that paying attention to cause-effect relationships in a text can help them understand why and how things happen. Have students read page 7 of *Amazing Armadillos.* Say, *The Spanish explorers thought the armadillo's skin looked like armor. This caused the explorers to name the animal* armadillo, *or little armored one. The armadillo's appearance was the cause and the Spanish name was the effect.*

Cause	Effect
The Spanish explorers think the armadillo's skin is like armor.	They name it "little armored one."

Practice/Apply After they have finished reading the book, ask students to identify the cause of the armadillo being named the official small mammal of Texas.

Reread to Clarify

Focus Strategy

Teach/Model Explain to students that good readers pay attention to whether they understand what is happening in a story and then reread to clarify anything that is confusing. Have students read pages 2–3 of *Amazing Armadillos.* Say, *When I read these pages, at first I am confused because they are about people, not animals. However, when I go back to the beginning and reread, I understand that the author is setting the scene by telling a legend. Rereading complicated passages will help you understand new words and ideas.* Remind students that if the cause or effect is unclear, they can also reread to clear up confusion.

Practice/Apply Tell students to apply the Reread to Clarify strategy whenever they come across a part of the text that seems confusing.

Build Vocabulary

Help students use context to understand key vocabulary. Have them answer each question by reading the words and sentences that surround the word.

eventually (p. 14)	What will *eventually* happen when you get older? **Possible response: I will drive a car.**
converse (p. 4)	Do you like to *converse* on the phone? **Possible response: yes**
continent (p. 14)	Which is the coldest *continent*? **Antarctica**
sphere (p. 11)	Which is a *sphere*, a ball or a block? **ball**
universe (p. 4)	Name something besides planets you find in the *universe*. **Possible response: stars**
homeward (p. 14)	When you head *homeward* after school, what do you do? **Possible response: play**

CM 175 **Independent Practice:** Vocabulary Practice on page 175

Read and Comprehend

Monitor Comprehension

Guide students as they read the text using the following questions:

Pages

2–6 What does the prefix *mis-* mean in a word such as *misbehaved*? ***mis-* means *not*, as in "not behaving"** PREFIXES, SUFFIXES, AND ROOTS

7–8 What does the author say an armadillo looks like? Why? **It looks like a little stone carving; Its skin is hard and gray.** REREAD TO CLARIFY (Focus Strategy)

9–11 Name three things an armadillo eats. **ants, worms, frogs** IMPORTANT DETAILS

What does scaring an armadillo cause it to do? **run away, try to look big and scary, scratch its opponent, or dig a quick hole** CAUSE-EFFECT (Focus Skill)

12–13 How is a baby armadillo's carapace different from an adult's? **A baby's is soft and pink, and an adult's is hard and brown or gray.** ANALYZE

14–16 List three states where armadillos are found. **Texas, Florida, Louisiana, Oklahoma, Kansas, Arkansas, Missouri, Georgia, South Carolina, Alabama** USE GRAPHIC AIDS

Think and Respond

Have students answer the questions on the inside back cover of the book.

1. **This story is organized by using chapters.**
2. **The armadillo's hard skin, or bony shell, protects it.**
3. **The main idea of this story is that armadillos are survivors.**
4. **Possible response: This story taught me that all creatures have their own ways to survive.**
5. **The tricksters' behavior was different because before they were embarrassed they disobeyed their elders and caused trouble, but after they were embarrassed they never caused trouble again.**
6. **Possible response: An interesting thing about the armadillo is it has been around for many years.**

Social Studies Connection Ask students to research another legend about how a particular animal came to be. Have students present their findings to the class.

School-Home Connection Ask students to complete a fact chart about armadillos before they leave for the day. Have students share their fact charts with a family member at home.

CM 176 **Independent Practice:** Comprehension Practice on page 176

Write

Writer's Craft: Word Choice

Write an Invitation Tell students that good writers choose colorful words, such as vivid verbs and specific nouns, to make their ideas clearer. Have students reread page 13 of *Amazing Armadillos.* Say, predators *and* paddles *are colorful, exact words.* Tell students to write an invitation for a trip to the zoo. Tell students to choose vivid words to describe details like which animals to see and what to do. Remind students that invitations include a heading, greeting, body with details, time and place, and signature.

Fluency

Oral Reading Fluency

CM 177

Assessment Use Oral Reading Fluency Practice page 177 to assess students' oral reading fluency.

Name ______________________

Amazing Armadillos
Vocabulary

▶ **Fill in each blank in the paragraph with the word that fits best.**

eventually	converse	continent
sphere	universe	homeward

Today in art class, my teacher told us we could draw whatever we wanted. At first, I was going to draw a picture of Europe, which is a **continent** we are learning about in social studies class. I **eventually** changed my mind and decided to draw the entire **universe**. It was fun because I got to draw the sun, stars, planets, and other objects that are shaped like a **sphere**. I can't wait to head **homeward**. I'm going to **converse** with my family about my drawing and tell everyone how much fun I had!

▶ **Imagine you are a zookeeper in charge of the armadillo exhibit. Using three vocabulary words, write four sentences telling the visitors at the zoo about the armadillos.**

Possible response: Welcome to the zoo. Armadillos live on the continent of North America. They also live in parts of South America. They armadillo cannot roll into a sphere to protect itself. If it needs to run away, it will probably just run homeward.

Vocabulary Practice p. 175

Name ______________________

Amazing Armadillos
Comprehension

▶ **Decide whether the following statements are True or False. If the statement is false, rewrite it to make it true.**

1. The bony shell of the armadillo is called the skeleton.
   False; The bony shell of the armadillo is called the carapace.
2. The armadillo has a keen sense of taste.
   False; The armadillo has a keen sense of smell.
3. Baby armadillos are called pups.
   True
4. The armadillo's eating habits cause it to be huge.
   False; The armadillo is not a huge animal. It grows to be about 2 ½ feet long and weighs 15 pounds.

▶ **What do scientists think caused armadillos to move north from South America?**

Scientists think armadillos were looking for food and got lost. They were not able to find their way home, so they dug burrows in North America and stayed there.

Comprehension Practice p. 176

Name ____

▶ **Fill in each blank in the paragraph with the word that fits best.**

eventually	**converse**	**continent**
sphere	**universe**	**homeward**

Today in art class, my teacher told us we could draw whatever we wanted. At first, I was going to draw a picture of Europe, which is a ____________________ we are learning about in social studies class. I ____________________ changed my mind and decided to draw the entire ____________________. It was fun because I got to draw the sun, stars, planets, and other objects that are shaped like a ____________________. I can't wait to head ____________________. I'm going to ____________________ with my family about my drawing and tell everyone how much fun I had!

▶ **Imagine you are a zookeeper in charge of the armadillo exhibit. Using three vocabulary words, write four sentences telling the visitors at the zoo about the armadillos.**

____

____

____

____

____

____

Name ______________________________

Decide whether the following statements are True or False. If the statement is false, rewrite it to make it true.

1. The bony shell of the armadillo is called the skeleton.

____

____

2. The armadillo has a keen sense of taste.

____

____

3. Baby armadillos are called pups.

____

____

4. The armadillo's eating habits cause it to be huge.

____

____

What do scientists think caused armadillos to move north from South America?

____

____

____

____

____

Name ______________________________

Once you see an armadillo, you will be able to understand the legend 13
better. An armadillo looks a little like a stone carving. It looks this way 27
because its skin is hard, like armor. Almost the entire body of the armadillo 41
is covered with this hard skin. This bony shell is called a *carapace*. 54

Spanish explorers gave the armadillo its name. They thought the 64
armadillo's skin looked like their own armor. In Spanish, *armadillo* means 75
"little armored one." 78

Scientists believe that armadillos were on Earth many years before 88
Spanish explorers discovered them. They may have lived 55 million years 99
ago. They may even be related to the glyptodont, an animal that lived at the 114
same time as the dinosaurs! 119

The armadillo's carapace is not one solid shell like a turtle's or snail's. In 133
the middle are nine bands. The bands are joined with skin. This allows the 147
armadillo to bend its body and move easily. 155

An armadillo has a long face with a long snout. 165

FLUENCY SCORE

Total Words Read Per Minute		________
Number of Errors	−	________
Number of Words Read Correctly		________

Our Moon

by Michael Burgan

SUMMARY The moon is our closest space neighbor. As it orbits the Earth, the moon seems to change shape as different parts of its surface reflect the sun. The moon's gravity affects ocean tides, and its surface is covered with craters and hills. Though scientists have studied the moon and astronauts have explored its surface, many questions about it remain unanswered.

- **Focus Skill:** Locate information
- **Focus Strategy:** Use Text Structure and Format
- **Vocabulary:** fluorescent, force, loops, nucleus, particles, solar wind

Differentiated Instruction

Paired Reading Organize students in pairs. Have them read the selection aloud to each other, alternating with each page. After students have finished reading, have them discuss in pairs a part of the story they found interesting.

Prepare to Read

Preview/Set Purpose Ask students what they know about the moon. Have them preview the book by examining the cover and title. Point out that this selection is nonfiction, a story that is written to teach us something. Ask students to predict what they will learn about the moon.

Focus Skill

Locate Information

Teach/Model Display a chart like the one below. Explain to students that making an outline as they read a text can help them locate information. Tell students to read pages 2–4. Say, *In the first section, we learn that the moon, our neighbor in the sky, circles Earth in an orbit and is the closest object to us in space. In the second section, we learn about why the moon's shape seems to change. Putting these facts in an outline will make it easier to locate information later.* Remind students that they also can use illustrations, headings, captions, and the table of contents to locate information in a text.

I. Our Neighbor in the Sky	II. Why Does the Moon's Shape Seem to Change?
A. moon circles Earth in an orbit B. closest object to us in space	A. goes through phases B. sun's light bounces off half the moon

Practice/Apply Have students make an outline of pages 6–8 after they read the story.

Focus Strategy

Use Text Structure and Format

Teach/Model Point out that good readers use text structure and format to understand what they read. Tell them that many nonfiction selections include headings, illustrations, and captions. Major ideas are organized under headings and captions give information about illustrations. Have students look at page 6. Say, *The heading is* What Is an Eclipse?. *This tells us that the main idea on these pages is the eclipse.*

Practice/Apply After reading, have students write a paragraph explaining how they used text structure and format to help them understand the selection.

Build Vocabulary

Help students use context to understand key vocabulary. Have them answer each question by reading the words and sentences that surround the word.

fluorescent (p. 13)	Name something *fluorescent*. **Possible response: light**
force (p. 10)	What is another word for *force*? **Possible response: power**
loops (not in text)	What shape are *loops*? **circles**
nucleus (p. 15)	Where is a *nucleus* located? **center**
particles (p. 9)	What is another name for *particles* of wood? **Possible response: sawdust**
solar wind (p. 16)	Where does *solar wind* come from? **the sun**

CM 181 **Independent Practice:** Vocabulary Practice page 181

Read and Comprehend

Monitor Comprehension

Guide students as they read the text, using the following questions:

Pages

2–5 How many phases of the moon are shown in the illustration? **seven** USE GRAPHIC AIDS (Focus Strategy)

6–7 Compare and contrast the two types of eclipses mentioned in the text. **lunar: Earth is exactly between sun and full moon; solar: moon passes between sun and Earth** COMPARE AND CONTRAST

8–11 Why can't we see the moon during the day? **The sun's light is much stronger than the moon's, so we can't see the moon's light.** CAUSE-EFFECT

Explain whether or not gravity exists on the moon. **Yes, because the moon's gravity is one-sixth that of the Earth's.** MAIN IDEAS AND DETAILS

12–14 Have you ever seen a face in the moon? What did it look like? **Possible response: Yes, it looks like the face of a snowman.** EXPRESS PERSONAL OPINION

What signs of life do you think the astronauts were looking for on the moon? **evidence of waste, water, or air** DRAW CONCLUSIONS

15–16 On which page do you learn about solar winds? **page 16** LOCATE INFORMATION (Focus Skill)

Think and Respond

Have students answer the questions on the inside back cover of the book.

1. **Each section after the first begins with a question.**
2. **The section titled "How Does the Moon Affect the Oceans?" talks about tides.**
3. **Possible response: People have so much interest in the moon because it is so different from Earth.**
4. **Possible response: The moon seems to change shape because as it moves around the Earth, we see different parts of the sunlit half.**
5. **Possible response: It is similar because there is gravity on both the moon and Earth. It is different because there is no life on the moon.**
6. **Possible response: I would like to visit the moon so I could see a different world.**

SCIENCE

Science Connection Have students make a list of any facts they know about space exploration.

School-Home Connection Have students write a paragraph about whether or not they would like to go to the moon.

CM 182 **Independent Practice:** Comprehension Practice page 182

Write

Writer's Craft: Word Choice

Write a Story Tell students that good writers use exact words to make ideas clear. Have students reread page 9. Say, Hazy *and* encircles *are exact words because they describe how the ring looks and what it does.* Tell students they will write a story about a trip to the moon. Explain to students that stories are written to entertain and have characters, a setting, and a plot. Remind students to use exact words in their stories.

Audience? classmates and teacher
Purpose for writing? to entertain
Setting?

Fluency

Oral Reading Fluency

CM 183

Assessment Use Oral Reading Fluency Practice page 183 to assess students' oral reading fluency.

Name ______________________

Our Moon
Vocabulary

▶ **Write the word that best completes each sentence.**

fluorescent	particles	loops
force	solar wind	nucleus

1. It was hard to see the road because of the **force** of the wind and rain.
2. I took my time and made perfect **loops** on my capital letters.
3. The **fluorescent** lights were bright and helped us find our way.
4. The center of a comet is called its **nucleus**.
5. Rain **particles** made me blink as I walked home from school.
6. **Solar wind** comes from the surface of the sun.

▶ **Create a poster that highlights the most interesting facts about the Moon, using at least three vocabulary words.**

Possible response: The moon is so bright it is fluorescent. The moon's gravity is a force that can pull the tides. The moon loops around the Earth.

▲ Vocabulary Practice p. 181

Name ______________________

Our Moon
Comprehension

▶ **Circle the best answer to the questions below.**

1. When does the moon have a reddish glow?
   - A during a solar eclipse
   - B very early in the morning
   - C right before the sun sets
   - Ⓓ during a lunar eclipse
2. The diagram on page 5 summarizes which of the following?
   - F causes of high and low tides during the summer months
   - G phases the sun goes through during the year
   - Ⓗ different phases of the moon
   - J causes of the hazy white ring around the moon
3. What happened first in the story *Our Moon*?
   - A We learned about the hazy white ring around the moon.
   - B We learned about the phases of the moon.
   - Ⓒ We learned how far away our moon is from Earth.
   - D We learned how the moon affects the oceans.
4. What causes the Northern Lights?
   - Ⓐ They occur when solar winds enter the Earth's atmosphere.
   - B The lights are created by magic.
   - C No one knows how they are made.
   - D They happen during a solar eclipse.

▶ **Answer the question below using complete sentences.**
Explain how the headings help you understand the text.

Possible response: The headings tell you the information you will read about in that section. The headings give you the main ideas of the selection. You can ask yourself questions from those headings, too.

▲ Comprehension Practice p. 182